Bureaucracy vs. Environment

Bureaucracy vs. Environment

The Environmental Costs of Bureaucratic Governance

John Baden and Richard L. Stroup, Editors

Ann Arbor

The University of Michigan Press

Copyright © by The University of Michigan 1981
All rights reserved
Published in the United States of America by
The University of Michigan Press and simultaneously
in Rexdale, Canada, by John Wiley & Sons Canada, Limited
Manufactured in the United States of America

1986   1985   1984   1983   5   4   3   2

Library of Congress Cataloging in Publication Data

Main entry under title:

Bureaucracy vs. environment.

   Bibliography: p.
   Includes index.
   1. Environmental policy—United States—Addresses, essays, lectures.   2. Environmental
policy—Economic aspects—United States—Addresses,
essays, lectures.   3. Conservation of natural
resources—United States—Addresses, essays,
lectures.   4. Bureaucracy—United States—Addresses, essays, lectures.   5. Self-interest—Addresses, essays, lectures.   I. Baden, John.
II. Stroup, Richard.
HC110.E5B88   1981   363.7′00973   81-10382
ISBN 0-472-10010-6   AACR2

Preface

The dawn of the environmental movement coincided with an increased skepticism of private property rights and the market. Many citizen activists blamed self-interest and the institutions that permit its expression for our environmental and natural resource crises. From there it was a short step to the conclusion that management by professional public "servants," or bureaucrats, would significantly ameliorate the problems identified in the celebrations accompanying Earth Day 1970.

Most simple answers are either wrong or incomplete. This is no exception. Bureaucrats, like most other people, are predominately self-interested. Given that an administrator's welfare tends to increase with increments in his budget, many of our resource administrators act as bureaucratic entrepreneurs. Unfortunately, unlike those in the private sector, these administrators are not accountable to a bottom line demarcating benefits and costs. Thus, the net "benefits" of many of their activities are strongly negative.

When the authority for decision making is not closely tied to the responsibility for outcomes, decision makers have few incentives to consider the full social costs of their actions. The articles in this book strongly suggest that this tendency generates problems in the public as well as in the private sector. Thus, we have significant problems in the public management of timber, range, water, and energy resources.

The key element in designing the solution lies in constructing institutions that provide incentives for individuals to take responsibility for the net impact of their decisions. The fundamental message of this book is that this principle can be applied in the public sector. The inability of bureaucrats to obtain a profit, in the conventional sense of that term, by no means guarantees that their decisions will be in the public interest. This book provides detailed cases of failures in the public sector and offers suggestions for policy reforms.

We would like to acknowledge those who have supported the work leading to this publication. The initial articles were written as part of the operating program of the Liberty Fund and were presented at a conference held at Big Sky, Montana. We would like to thank all those who participated in the conference, especially those who took the time to rework their articles for inclusion in this volume. We would also like to thank JoAnn McDonald, the Center's administrative secretary, and Marianne Keddington, Center editor, for helping to put the manuscript into final form. The Scaife Family Charitable Trust, the Earhart Foundation, the Samuel Roberts Noble Foundation, the Murdock Trust, and the AMAX Foundation contributed substantially to the final preparation of the manuscript. These organizations provided financial support for us and our staff so that we could complete the manuscript. Although we are extremely grateful for this support, it is our hope that the primary beneficiaries, now and in the future, will be the American public.

John Baden
Richard L. Stroup

Contents

Introduction
John Baden and Richard L. Stroup   1

The Federal Treasury as a Common Pool Resource
and the Development of a Predatory Bureaucracy
Rodney D. Fort and John Baden   9

Property Rights as a Common Pool Resource
Terry L. Anderson and Peter J. Hill   22

Social Costs, Incentive Structures, and
Environmental Policies
Lloyd D. Orr   46

U.S. Natural Gas Policy: An Autopsy
Ernst R. Habicht, Jr.   64

The Policy-Induced Demand for Coal Gasification
Richard L. Stroup   77

The Navajo and Too Many Sheep: Overgrazing
on the Reservation
Gary D. Libecap and Ronald N. Johnson   87

Dams and Disasters: The Social Problems
of Water Development Policies
Bernard Shanks   108

A Perspective on BLM Grazing Policy
Sabine Kremp   124

Chained to the Bottom
Ronald M. Lanner   154

An Institutional Dinosaur with an Ace: Or,
How to Piddle Away Public Timber Wealth and
Foul the Environment in the Process
  Barney Dowdle   170

Compounding Clear-cuts: The Social Failures
of Public Timber Management in the Rockies
  William F. Hyde   186

Transgenerational Equity and Natural Resources:
Or, Too Bad We Don't Have Coal Rangers
  John Baden and Richard L. Stroup   203

The Environmental Costs of Bureaucratic
Governance: Theory and Cases
  M. Bruce Johnson   217

Selected Bibliography   225

Contributors   229

Index   231

Introduction

by John Baden and Richard L. Stroup

The essays in this volume share a theme. Although this theme is important, clear, and compellingly valid in a large and varied set of cases, it is only slowly becoming understood. Specifically, we are increasingly convinced that both the environmental and the economic costs of bureaucratic management of natural resources are excessively and unnecessarily high. These social costs are generated by perverse institutional structures that give authority to those who do not bear responsibility for the consequences of their actions.

There is a substantial body of literature in economics and political science—and especially in their conjunction—that predicts the implications of analogous situations in the private sector. While explanations of failure in the private sector are generally understood and have been incorporated into syllabi for a least a generation, the public sector has only recently received comparable attention.[1] The primary social value of this collection will be its contribution toward explicating the costs, especially the environmental costs, of bureaucrats holding authority that is buffered from responsibility.

A major implication of these essays may be startling to some but becomes clear when carefully considered: there is a potential natural coalition made up of environmentalists and those oriented toward free market economics. William Reilly, president of the Conservation Foundation, made this point emphatically in a *Wall Street Journal* editorial of August 31, 1979:

> The President's energy proposals should be opposed vigorously by both environmentalists and business leaders. This may seem an "unholy alliance"—unrealistic and unworkable. But I ask both the business and environmental communities to consider the following. . . . [Under Carter's energy program] we would have the

dubious distinction of buying heavily subsidized synthetic oil at a price higher than OPEC would likely charge for the natural product. . . .

Accept for a moment my organization's viewpoint that a massive synfuels program is bad for the environment, and let us concentrate on how it is bad for business and industry. The President has suggested a program that removes a vital part of the American economy from the normal controls of competitive free-market tests. . . . Pricing that reflects the full costs of energy consumed is fully consistent with environmentalists' view of the world: that we can have economic growth and improved productivity with fewer investments of natural resources. . . . Cost effectiveness is the test, not some predetermined "synfuels or bust" strategy.

I believe it makes sense for important interests outside government to cooperate in confronting the energy peril. Of course there is much that separates business and environmentalists. But cooperation is possible even without consensus on every detail of energy policy.

Related points in this volume are made by Dr. Ernst Habicht, formerly a staff scientist with the Environmental Defense Fund, and Dr. Bernard Shanks, member of the board of the Wilderness Society. The social costs of ignoring their message that bureaucratic government is expensive may be high indeed.

In the early days of the environmental movement it was generally accepted that self-interest and the profit motive fostered environmental destruction. Rather than identifying the cause as poorly defined and defended property rights, i.e., political and institutional failure, the market system was generally blamed. Thus, a logical next step was to rely more and more on bureaucratic management. Since bureaucrats do not obtain profits, the underlying motivation for environmental destruction would be absent. Reality, however, intrudes. Bureaucrats, like others, respond to incentives. Self-interest tends to dominate considerations of environmental quality and, *even in the absence of profits,* often produces environmental costs.

During the decade since Earth Day, a large and growing number of articles, technical reports, and books have developed the theme that large social benefits can be generated through increased governmental action, management, ownership, and control. This is especially true in the area of natural resource management. A strong distrust of private property as an institution coupled with increasing demands for scarce natural resources and greater environmental quality has led to more and stronger calls for collective management to supplant or control what is viewed as the rapacious private exploitation of nature.

Privately held property rights in natural resources are increasingly attacked. This is partly due to concern over what many perceive to be an imbalance in the distribution of benefits from nature's bounty. It is also due to what is thought to be irresponsible stewardship of natural resources in the pursuit of profits. Market failure is increasingly noted. Negative externalities, which are costs accruing to those other than the decision maker, are given as justification for governmental control. The rule of willing consent is relaxed, and collective decision making is imposed.

The market, based on the willing consent of individuals and operating through the mechanism of prices representing condensed information and incentives, tends to move resources to the most highly valued uses. When transaction costs are negligible and property rights clear and readily enforceable, the market will, given any existing distribution of income, provide the socially optimal production of goods and services. Unfortunately, when dealing with some natural resources, there are only very imperfect property rights, as witness clean air and clean water. Resources such as these tend to be underpriced. As a result, the production process generates not only goods but also bads in the form of negative externalities. Because environmental goods tend to be public goods and common pool resources, the private market will not efficiently utilize them. In principle, injured individuals could collect damages through the courts, using nuisance laws. But transaction costs among a large number of disparate individuals are extremely high. Each tends to be a "free rider," not joining in legal action. Thus, to move toward the optimum allocation, the rule of willing consent is relaxed, and governmental mandates are imposed. Hence, great social benefits can *potentially* be generated via the imposition of governmental regulation and control.

There is, however, a set of environmental quality issues that has been relatively neglected by resource scholars. They are the subject of this collection. These issues involve reductions in environmental quality *generated* by the positive action of governmental agencies. It is easy for most of us to understand why profit-seeking individuals or firms will engage in environmentally destructive practices when the private cost to them of doing so is far below the social cost, due to poorly defined property rights. Fewer people, however, have an intuitive understanding of why bureaucrats, individuals whose salaries bear little or no relationship to profit, should engage in analogous behavior.

The root cause of failure in the collective arena is the same as in the private: authority and responsibility are separated. In the private sector, this occurs when property rights are not clearly established or enforceable so that, for example, the smelter owner uses the air resource for free garbage (SO_2) removal and is not held responsible. He captures benefits but not costs. Similarly—but much more frequently —in the public sector, the individual with the authority to order an action does not bear certain important costs of that action. When the Federal Power Commission keeps natural gas wellhead prices low, commissioners are applauded for protecting gas buyers. They are not condemned for causing the environmental damages due to the extra electricity production, the building of a multibillion dollar Alaskan gas pipeline, or environmentally and economically expensive coal conversion to natural gas. Even if they were blamed, the cost to them would not approach the sum of the total costs to society. As Tullock has noted, better public decisions really are public goods.[2]

In general, public sector activity suffers from what Hardin has labeled, in a nongovernmental setting, the tragedy of the commons. Everyone's property is no one's property. The public purse and public authority, like common pastures, are overutilized for the benefit of the few at the cost of the many. Efficient management is elusive and hard to recognize when users do not pay, and it is seldom attained. Further, a full and equal sharing of public benefits and costs is impossible even with one-man-one-vote democracy. A lucky few will enjoy the water from federally financed water projects, artificially cheap natural gas, or the extra forage produced by the chaining of piñon-juniper stands. Yet all incur the costs. The ability to influence government is probably no more (and arguably less) equally distributed than the money income that can influence market decisions. Yet, when concentrated "special" interest groups are found to wield power in obtaining large governmental benefits, specific people are usually seen as the cause of the problem. We argue instead that the system's *structure,* not evil individuals, can best explain such problems. The same frontier white men who nearly wiped out the buffalo posed no threat to the valuable beef cattle raised on the western range. Similarly, better institutional arrangements can better channel the efforts of imperfect men, the vast majority of whom could never be considered seriously for sainthood.

To foster understanding of the essays that follow, we intend:

1. to explain, using commonsense elements of the property rights paradigm and public choice theory, why we should not *expect* collective management to provide either careful stewardship *or* a balanced reflection of public desires;

2. to explain why this logic reflects the real world, so that collective management, usually thought to be the solution, is in fact often the problem; and

3. to provide the analytical foundations for developing private institutions and revising public institutions to foster efficient and noncoercive stewardship of natural environments.

The Framework of Analysis

In predicting or analyzing public sector behavior, the appropriate unit of analysis is the individual decision maker. We must also recognize that we operate in a representative, not a town-meeting, democracy. At the national level, politicians and bureaucrats are in control, and the average citizen knows little in detail about most government operations. In most of the cases that follow (and probably in most important cases), the key decision maker is the professional public servant, the bureaucrat.

Bureaucrats, like most other people, are largely self-interested. Like the rest of us, they will sometimes act altruistically to advance the public interest. In most work-related situations, however, a bureaucrat will act to improve his own welfare. The components of a bureaucrat's welfare include salary, relative position in agency, discretionary control over budget, perquisites of office, and work place amenities. The nature of the world happens to be such that the components of bureaucrats' welfare are improved when the agency is growing. Conversely, reduction of their welfare tends to be associated with the contraction of an agency. Thus, decision makers face strong incentives to continually expand the scope of their agency's activities.[3]

Unlike their counterparts in the private sector, the expansionary activities undertaken by agencies need not meet the reality check imposed by the requirement of generating value that exceeds cost. Because their budget is derived from taxes rather than the sale of products, they have incentives to ignore or exaggerate the economic efficiency of the projects that they propose, sponsor, and administer. Within this context the creative bureaucratic entrepreneur can identify a specific clientele that stands to benefit from a proposed action. In exchange for strong political support for his agency, he can finance this action with a subsidy from the general taxpayer. Hence, from the perspective of the bureaucrat, the tax base becomes a common pool resource ripe for exploitation. In contrast, the private individual or firm will not intentionally over-produce goods when the social value of the inputs is accurately reflected in the price of these inputs. The bureaucrat can ignore the foregone value of inputs. He will over-produce dams, grass, timber, and a host of other goods because the resources used in the production-tax money (or timber land) are tapped from a common pool.

In a representative democracy, bureaucratic discretion and its abuse are possible because of voters' rational ignorance.[4] Sometimes mistaken for simple apathy, the average voter's lack of diligence in learning about and trying to control agency behavior is easily explained by the lack of individual control and the *opportunity cost* of obtaining information. A given resource management activity may mean far more to a voter than where he buys his gasoline. Yet all the benefits of knowing where to buy gas are his and he controls the decision. Whether piñon-juniper stands will be chained is normally beyond his control. His learning time buys him more in shopping for gas than in learning about Bureau of Land Management or Forest Service policy. He rationally is ignorant about most important policy issues. Those with concentrated interests in a particular resource are not. Their marginal influence typically controls.

Most, if not all, of the environmentally destructive practices discussed in the following essays *would not occur if the agencies were required to meet the standards of economic efficiency.* In effect, the general taxpayer often finds himself subsidizing the destruction of the environment while making transfer payments to bureaucrats and highly specific special interests. There are huge social profits to be made by explicating and advertising this situation and constraining the activities of the agencies discussed.

Those who stand to benefit from constraints on excessive governmental action are numerous indeed. Obviously, the general taxpayer benefits if the government operates in a manner consistent with economic efficiency. Environmentalists should welcome a reduction in governmental programs that fail to meet the tests of economic efficiency and are demonstrably destructive of the environment. Such programs convert relatively pristine environments into chained, terraced, dammed, flooded, channelized, and similarly disturbed areas. It is hoped that most Americans would agree that we should engage in development when the net benefits exceed the costs but that we should stop *subsidizing* the destruction of nature. A third category of those likely to support careful analysis of governmental actions are those who view freedom as a scarce and valuable resource and who realize that the growth of government, whatever the benefits, constitutes an extraordinarily serious, pervasive, and unavoidable threat to that resource.

The groups discussed above are not mutually exclusive. All of us are taxpayers, many of us are environmentalists, and some of us set a very high value on freedom. The conjunction is not insignificant, and we hope that the number is growing. One of our colleagues has likened a land management agency to the *Titanic* steaming through the night. The selections in this volume serve as warnings to the agencies that icebergs lie ahead. We hope they will heed these warnings. If they do not, the public is likely to raise the decibels of the message. The cases are strong; the evidence is clear. The public, we feel, is increasingly receptive to the kind of analysis presented here and illustrated in the following chapters.

Analytical Foundations for Better Resource Management

The property rights paradigm helps us to pinpoint the cause of environmental problems in a market setting: resource prices are distorted when property rights to a resource are not enforced. Those with the authority to increase social welfare are not forced to be responsible for their actions. We should not expect environmental problems to be solved automatically when we give up the rule of willing consent and opt for collective action. Indeed, we are likely to create problems.

Two steps seem likely candidates in the search for systematic improvement. The first is to recognize the problem: the separation of authority from responsibility, all too prevalent in the private sector, is the norm in the public sector. What efficiency experts might call lack of accountability results. The second step follows from the first: we must accept an imperfect solution; market failures do not automatically imply that collective action is better.[5] An imperfect market may actually be the best available alternative.

The potential payoffs of doing more work on the political economy of managing natural environments are indeed large. The essays that follow represent some of this work. The problem has been identified in this volume. It has yet to be fully solved.

Notes

1. For a survey of this material, see J. D. Gwartney and R. L. Stroup, *Economics: Private and Public Choice,* 2d ed. (New York: Academic Press, 1980), chaps. 4, 31, and 32.

2. For a discussion of the danger of assuming the benevolence of government, see G. Tullock, "Social Costs of Reducing Social Costs," in *Managing the Commons,* ed. Garrett Hardin and John Baden (San Francisco: Freeman, 1977), pp. 147–56.

3. Variations of this theme are presented in T. E. Borcherding, ed., *Budgets and Bureaucrats: The Sources of Government Growth* (Durham, N.C.: Duke University Press, 1977); R. B. McKenzie and G. Tullock, *Modern Political Economy* (New York: McGraw-Hill, 1978); and W. A. Niskanen, Jr., *Bureaucracy and Representative Government* (Chicago: Aldine-Atherton, 1971).

4. Rational ignorance and other problems in collective decision making are discussed in Gwartney and Stroup, *Economics,* and in McKenzie and Tullock, *Modern Political Economy.*

5. The opposite is also true: an imperfection in collective management should not automatically cause us to avoid governmental action. The grass is *not* always greener. The fate of the environment under a collectivist system has recently been explored in three studies: Fred Singleton, ed., *Environmental Misuse in the Soviet Union* (New York: Praeger, 1976); Phillip R. Pryde, *Conservation in the Soviet Union* (New York: Cambridge University Press, 1972); and Marshall I. Goldman, *The Spoils of Progress: Environmental Pollution in the Soviet Union* (Cambridge, Mass.: MIT Press, 1972). These books were reviewed by Robert J. Smith in *Policy Review,* Fall 1977.

The Federal Treasury as a Common Pool Resource and the Development of a Predatory Bureaucracy

by Rodney D. Fort and John Baden

Pessimism over the prospect of reducing the size and scope of government is pervasive. As Ralph Winter notes in *Regulation,* part of the basis for this pessimism is that, as government grows, elections become less and less relevant to outcomes.[1] In this immobilizing ambiance, government grows apace with antigovernment sentiment. The general purpose of this paper is to provide an important reason for this paradox.

We contend that elections fail to control government size and growth because of specific failures in the representative system. One major failure has been the increase in political activities within bureaucracies. This shift away from the representative arena is a result of placing increased responsibility for administering the "transfer society" in the hands of the bureaucracy. At both the level of individual interaction with agencies and the level of interagency interaction, the pervasive result of government growth, as distinguished from absolute size, is manifest. It is time to suggest plausible modifications.

Government Size: Divergence of Individual and Social Cost

Economic interaction in the United States has occurred primarily within the market system. Adam Smith, in the eighteenth century, prophetically saw the ability of unfettered markets to facilitate mutu-

The authors wish to thank the Scaife Family Charitable Trust for support in preparing this article. A version of the paper has appeared in *Policy Review* 11 (Winter 1980): 69–81. Used by permission.

ally beneficial transactions in complex settings, what he called the "invisible hand." Alas, for all their virtues, markets sometimes fail.

The invisible hand acts decisively on existing resource information as reflected by the market price of resources. Goods are supplied according to the additional benefits and costs of doing so; output is constrained by the price of inputs. If the total social costs of inputs (i.e., all costs including those external to the producer) are not reflected by the actual market price, then the price of the product does not reflect the total value of what society gave up in order to have that good. In other words, if externalities are present, the market price does not account for the total opportunity cost of the good.

For example, the notion that "air is free" is taken to heart by the invisible hand as it allocates resource use. Air is treated by individual producers as a cost-free dump for by-products of production. In general, where property rights to resources are poorly defined and enforced, perceived private costs (or benefits) and true social costs (or benefits) diverge. The output that results under such a scheme is nonoptimal.

Other potentially important market failures also include the existence of (*a*) market power and (*b*) public goods for much the same reason—divergence of perceived individual and social costs or benefits. In the case of market power (monopoly, duopoly, and oligopoly are common references), the *social* value of additional units of output is greater than the cost to the producer, but the benefits of these additional units to the *producer* are less than their cost. Unless we have a perfectly discriminating monopolist, the result is both restricted output and a higher price than is socially optimal. To economists, this is the familiar problem of the lost "welfare triangle."

In the case of public goods[2] (the textbook example is national defense), the total cost of providing the good may be greater than the benefit to any individual, while the benefits of the good summed over all of the beneficiaries in society are greater than the total cost of provision. This total cost and individual benefit divergence, plus the problem of individuals holding out with the hope that others will provide the good so that they need not contribute (the "free rider" problem stressed initially by Olson), also results in a socially suboptimal amount of public goods provision in the absence of coercion.[3] In all of these instances, the individual decision process results in outcomes that are not as they would be if total social costs were taken into account.

Ironically, government acts to maximize public interest, and it fails for the same reasons markets sometimes fail. The costs perceived by individual decision makers do not accurately reflect the total social costs of their decisions; government fails because perceived individual costs and true social costs diverge. Hence, we expect government output to be socially nonoptimal and government itself to be too big. Since this paper focuses on bureaucracy, here is an example.[4]

The national forests of the Rocky Mountain states are much less productive for growing trees than the national forests of Oregon and Washington. The greater value of most Rocky Mountain forest lands is for recreation. Timber practices on these lands, however, have often impaired these relatively higher values. Clear-cutting is a prime example. Much of this harvested timber literally has negative value *as timber.* So, in this less productive timber area, the true social costs of timber practices, e.g., the opportunity cost of the diminished value of camping trips to a clear-cut area, were often discounted by forest managers. Since some social costs were not considered, the resulting timber output is socially nonoptimal.

Government Growth: Dispersed Costs and the Transfer Society

The reason that government grows can be traced to two ideas: (*a*) government will succeed in overcoming the problems that cause markets to fail, and (*b*) investments in influencing governmental decisions may be profitable. While the first belief lacks compelling corroborating evidence, it continues to flourish. The second, unfortunately, is substantially correct under the existing institutional structure. While it is the divergence between the costs faced by the decision maker and the total social cost resulting from his decision that makes government too big, it is the increased propensity of government decision makers to generate concentrated benefits to special interest groups that leads to government growth. The government's ability to disperse the costs of this benefit generation to all taxpayers (or even to future elections and generations) could be the "hole in the dike" that allows these increased transfers.

Anderson and Hill provide substantial evidence in support of the argument that the United States has recently become and is continuing to be an ever-larger "transfer society."

The early American experience was one in which transfer activity was very limited and productive activity was encouraged. But

> because of the alterations in the institutional framework or the rules under which economic activity takes place, that situation has reversed. We are now a society in which transfer activity is encouraged at the expense of productive activity.[5]

Their argument hinges on the idea that the continuous altering of social rules, i.e., the court's interpretation of the Constitution, has favored transfer-seeking activities. Since transfer activities occur in the political arena, the result of continuous rule changing in favor of transfer seeking has resulted in government growth that will continue as long as the rules are so altered.

More support for a bureaucratic agency can be generated by increasing benefits selectively than by reducing costs generally. It is bureaucratically profitable to cultivate a concentrated group of beneficiaries. In the clear-cutting example, this group was the commercial timber interests. We could not overstress the importance of understanding that government growth is the result of rational behavior. It is patterned and predictable. We can understand its cause. Favor-seeking and favor-provision are marginally beneficial! Anderson and Hill's transfer society flourishes. Further, as government grows, more individuals learn to play and have a stake in the game. Few realize, or are forced to account for, the social costs of personally beneficial programs. The costs are distributed among all taxpayers. But individuals do realize that self-denial will not be reciprocated; giving up their benefits does not mean that they do not have to pay for programs beneficial to others!

The propensity of government to increase transfer activities requires an administrative force to carry out the transfers. It is this shifted focus toward bureaucratic administration that leads us to an examination of bureaucrats and the treasury commons. Nonreciprocated self-denial is as operative among agencies as it is among their supposed beneficiaries.

Bureaucrats and the Treasury Commons

The term "bureaucrat" will be used to identify the decision makers in administrative governmental agencies. Typically, these individuals are public servants whose public actions are presumed to be in the public interest. As the outcomes of bureaucratic activity are closely scrutinized, it becomes apparent to many observers that our public servants often produce results that can only be regarded as in the interest of some concentrated groups and in the interest of the bureaucrat himself. When this conclusion is reached, cynical condemnation of public servants often follows. While this may provide psychic unguent, it retards remedial action by diverting attention from the causes of perverse bureaucratic outcomes. Focusing on "bad" bureaucrats clouds the issue. Bureaucrats of even the purest intentions cannot be expected to produce results consistent with the welfare of their wards if by so doing they harm their own professional welfare.[6] Since bureaucratic outcomes are frequently in violation of their public interest functional designs and bad intentions cannot be assumed, we contend that the incentive structures faced by bureaucrats are perverse, and incentives are responsible for outcomes. There has been little talk of rigging the rules of the bureaucratic game, by manipulating costs and rewards to reduce this curious asymmetry between designed purpose and actual outcome. This section provides the behavioral basis upon which an ameliorative model of the predatory bureaucracy can be developed.

Garrett Hardin defines common pool resources as a lack of exclusive ownership. As a result, demands on the resource can be expected to exceed its capacity to meet them. Some positive net utility is individually perceived from all captured portions of the commons; the individual receives all of the benefits from captured portions while the costs of his actions are dispersed over the community of users in the form of lost capture opportunities. Rationality dictates ever-increasing capture as sensible to each individual. Hardin concludes that when all users pursue their own interests in a commons the outcome is tragic to the productivity of the commons and, hence, to the users.[7]

In justifying treatment of the treasury as a commons, it must be recognized that, in this case, the community of users is the entire federal bureaucracy. While "ownership" may be a questionable expression in terms of semantics, it has explanatory advantage. If anyone can be said to own that portion of the treasury allocated to bureaucratic purposes, it is the Appropriations Committee of Congress. However, as shall be argued, the effective ability of Congress to exclude bureaucrats from the treasury is weak. In an important sense, this brings into question whether or not exclusive rights over the treasury actually belong to Congress.

At first glance it appears that since agency budgets are determined by the "owner" of the treasury (Congress) *no* access rights are held by the agencies. One could then conclude that agencies exist as a result of a careful review of their relative success at fulfilling designed purposes. Recurring evidence of the divergence between designed intent and actual outcomes of the bureaucratic process casts serious doubt on both premise and conclusion. While technically no rights to existence reside within any agency, their power to extort the means of existence from Congress is great. Rourke's examination of the cultivation of powerful clientele groups and specialization of functions[8] plus McKenzie and Tullock's model of the monopsonist-monopolist relationship of government to agency[9] support the view that the exclusive rights of the Appropriations Committee over the treasury are very weak indeed. Real world examples of this political clout reinforce the view that agencies are actually powerful enough to claim rights to existence and, thus, to the treasury. When all agencies can do so, the treasury is essentially fair game to all—the essential ingredient to the existence of a common pool situation.

Note that rights to existence can become independent of an agency's designed function. Cultivation and expansion of the means of existence become overriding concerns. This diversion of resources to the continued existence of the agency is a result of institutionalized incentives, not some inherent malevolence found in public servants.

A second characteristic of the treasury that justifies its analysis as a commons is that demands on the resource exceed its ability to supply them. At a given point in time, the treasury is finite while wants are not. If this were not the case, there would be no need for Appropriations Committee hearings to decide on the distribution of the budget. Some observers may contend that supply is adequate or exceeds reasonable demands (as specified by the observer's subjective criteria), but all acknowledge that agencies compete for budget and, therefore, budget is scarce. The ability of bureaucrats to pressure for an ever larger budget, as their incentive structure suggests they will, is a different matter taken up later in this paper.

Do bureaucrats behave as Hardin's model predicts they will? For a commons to be exploited, individual calculus must dictate that the pursuit of self-maximization has primary importance. In the case of the career bureaucrat, the maximization of personal welfare becomes inextricable from the maximization of his agency's welfare, particularly its budget. Agency welfare measures include expanding its employment capacity and its scope of activities, both of which entail increased funding. An important measure of the bureaucrat's professional welfare is his discretion over the allocation of agency resources.

Certain reservations are in order regarding the idea that bureaucrats are self-maximizers. Risk aversion and the long-run career orientation of bureaucrats may categorize them as "satisficers."[10] That is, bureaucrats aim at a "satisfactory" rate of agency and personal interest increases by strategically and carefully applying their continually cultivated treasury extortion factors. Bureaucrats maximize their self-interest, or utility function, subject to the constraints imposed by the incentive structure in which they operate. While this is not necessarily the same as maximizing profits,[11] it is sufficient for our purpose of stating that self-interest maximization is an important ingredient in the bureaucratic arena. Self-interest is also the driving force behind the tragedy of common pool resource destruction.

The Tragedy of the Treasury Commons

Nonexclusive ownership and self-maximizing behavior exist in sufficient quantity to justify labeling that portion of the treasury allocated to bureaucratic budgets as a commons. Accepting the treasury as a common pool resource allows us to apply the logic of Hardin's tragedy of the commons model. Seeking to maximize his gain, each bureaucrat realizes that he has clear access to the treasury. He can be seen as asking, "What is the gain to my organization (hence, to me) of capturing another increment of the treasury?" All of the gain would go to finance his agency's activities while the costs of his capture are spread among the entire community of bureaucrats in terms of lost capture opportunities. All bureaucrats realize that the same calculus holds for them and that it is rational for each to capture additional increments of the treasury. Each bureaucrat must find ways to increase his agency's magnitude and scope. While Hardin deals with the analogous human-ecosystem interactions (his example is of herdsmen on a common pasture), human-human interactions are equivalent. Some ensuing implications of the tragedy of the treasury commons deserve mention.

First, negative spillovers in terms of lost opportunities to the rest of the bureaucratic community due to the independent or "free" actions of individual bureaucrats would suggest a fervent sense of competition for budget capture. The ameliorative of interagency predation is rendered impotent when it is made easier to receive funding by increasing the size of the pie; in other words, the total budget simply grows. A lesson well-learned in the commons is that self-denial will not be reciprocated. Given that bureaucrats operate within the treasury commons, should this lesson not be as apparent to them?

Second, in the absence of signals such as prices and consumer preference and in response to increasing pressures to justify higher expenditures, investment in agencies' programs at nonoptimal times and in nonoptimal amounts is to be expected (and is evident). It is a significant failure of government that often the discount rate of elected officials and bureaucrats is greater than that reflected by intertemporal decisions made in the marketplace. While government is often criticized for being too concerned with future generations, it is also true that costs of government can be and are sloughed off onto future elections and even unborn generations through legislative and bureaucratic shortsightedness.

Third, the chance to attain some spillover benefits to the community of bureaucratic users may be realized. By introducing some collusive mechanism, it would be possible for bureaucratic leaders to magnify their individual impacts in tapping the treasury through mutually beneficial arrangement. While community restraint through "mutual coercion mutually agreed upon" can avert the tragedy of the commons,[12] cooperation among agencies has the potential to further intensify demands on the commons.

Fourth, it is in the interest of all bureaucratic community members to maximize the size of the pie from which their budget captures arise. One infers from the logic of the commons that treatment of the commons to the best ends of the bureaucratic community would favor increasing the *absolute size* of the commons rather than utilizing it in a cost-efficient manner. Thus far, the ability of actors within the bureaucratic incentive structure to divert ever-increasing amounts of society's productive capacity to their own ends has proven formidable. The bureaucratic process allows the generation of concentrated benefits for special interest groups. Simultaneously, the costs of providing such benefits can be dispersed and hidden. Costs of regulation, costs of inflation,[13] and the ability to bestow costs on future generations are all examples.

One must realize that it is the willingness of individuals to pay taxes that ultimately limits the treasury. It is the taxpayer's income that, unwittingly or not, actually falls prey to the institutional commons. Losses inherent to the tragedy of the treasury commons are borne by all society members in the form of lost control over productive resources and a relaxation of the rule of willing consent. As more decisions over any individual's resources are made without his consent, the greater is the chance that a decision will be unsatisfactory to that individual. This will affect each of us as the scope and magnitude of the public sector increase. By most criteria, then, these losses are in an increasingly scarce currency—freedom in everyday life. As Hardin so aptly states, "Freedom in the commons brings ruin to all." Pursuit of bureaucratic self-interest in the treasury commons is predicted to bring tragedy when all bureaucrats, according to their incentive structure, set such a course.

Development of a Predatory Bureaucracy

Several generations of economists and others interested in policy analysis have noted that a substantial proportion of legislation has socially wasteful results. At a time when many resources are perceived as becoming scarce, many people are disturbed by this waste. Further, most of these economists complain that their analyses are noted and then ignored—or merely ignored—in the political sector. Except for those lost in the wonder of their analytical creativity, a reaction of hurt resignation is fully expected. The economist applying cost-benefit analysis to federal projects becomes a contemporary Sisyphus. Rather than rolling stones endlessly up slopes never to reach the top, they unfold printouts before committees whose lenses have been ground by special interests. It is not that the analysts' products are necessarily flawed, it is merely that their political environment is unreceptive. As a result, the potential utility of their product is unrealized. Neither good intentions nor good products will suffice. Too many interests have too large an incentive to ignore the output.

Decisions are made on the basis of information and incentives. In this case, there is little incentive to utilize the information available. There is at least one obvious institutional solution to the problem: the creation of a "predatory bureaucracy."

The literature on bureaucratic pathology is voluminous and growing. In its traditional form it exists in public administration, political science, and sociology. Recent advances, however, have come largely from applying economic logic to the area. The conclusions reached in each of these areas remain fairly consistent with the following: bureaucrats operate to increase their discretionary control over resources. In sum, they operate to expand their budgets.

Writing on the Civil Service Reform Act signed in October 1978, Stephen Miller, a resident fellow at the American Enterprise Institute, notes that

> Of course the new law will not solve the problem of bureaucracy. Nothing really will. Bureaucracy is less a problem than a disease of modern civilization, one that can be treated but not cured. Like air pollution, one can't do away with it altogether.[14]

He goes on

> Given the dynamics of the Washington establishment, it is extremely difficult to eliminate ongoing programs. In order to do so, a counter-constituency has to be organized, one that is strongly opposed to a particular program. But it is hard to organize people to oppose something unless they have compelling reasons to do so. People are against inflation, bureaucracy, unemployment, or abortion; rarely are they against a particular federal program. Once a program—or a set of programs organized under the rubric of an agency—is put into motion, it tends not only to stay in motion but also to stay on the same course, not changing its way of doing things until scandal throws it off course.[15]

We have indicated why governmental budgets have a strong propensity to grow and noted a generalized recognition of a near cancerous bureaucratic pathology. Further, we have stated a rather obvious but fundamental belief that decisions are made on the basis of information and incentives. Finally, we have stressed that the current institutional setting fails to provide those incentives requisite to successful efforts at budgetary reduction. Yet, there are grounds for cautious optimism. Clearly there exists, at least in principle, a potential for institutional modifications that will ameliorate the problem of growth in the governmental sector. The fundamental issue is one of designing an institutional environment that will provide incentives to use information erosive to agency budgets.

A predator is an animal (or occasionally a plant) that captures and extracts its sustenance from other animals. Could this mode of existence be replicated and introduced in a bureaucratic environment to provide a negative feedback to bureaucratic growth? Conceptually, the answer is yes, but objections should be anticipated. First, what is the structure?

Assume that the Bureau of Budgetary Control (BBC) is initiated as a one-sided agency. It is admittedly designed to represent *one* position and to serve as an advocate of one fundamental goal: budgetary reductions. The design problems become (*a*) providing incentives to perform and (*b*) structuring incentives for this bureau to predate on those budgetary items whose social costs promise to swamp social benefits.

Further, assume that this agency is established with a *one-time appropriation that will carry it for two years only.* This constraint is critical. It is at this point that we harness the fundamental pathology of bureaucracies, that glaciallike propensity toward perpetuation and growth, for social benefit. Continual funding, and hence survival and growth, will depend on predation of other agencies' budgetary requests. As Miller stated, "It is hard to organize people to oppose something unless they have compelling reasons to do so." This strategy provides compelling opportunities for the proposed Bureau of Budgetary Control.

Assume that the Bureau of Reclamation requests $250 million to rebuild Teton Dam, again primarily as a flood control project. A number of local farmers who grow subsidized grain and sugar beets support the project. While it is obvious that if some dams fail to fail we may be overbuilding dams (i.e., the safety factors may be too high on the margin), it is also obvious that this particular project is of extremely dubious value on net. Hence, the BBC would marshall evidence in direct opposition to the testimony developed by the Bureau of Reclamation and its clientele. Of course, the BBC would also have strong incentives to develop client groups.

If Congress rejects the proposal, then two budgetary transfers are made. First, the BBC receives one percent of the requested budgetary item. Second, the proposing agency, in this case the Bureau of Reclamation, suffers a budget cut of one percent of the project's proposed operating costs. (These figures are strictly arbitrary and are likely to benefit from adjustment based on experience.)

The major advantage of this system is that it counters the problem of legislation that concentrates benefits while diffusing costs. Further, it builds into the appropriation process a spokesman for the public interest—more important, a spokesman who does good while doing well. By employing this system we rely on self-interest to advance the public interest. There are, of course, a few technical problems with this proposal, but they are likely to be minor when compared with the benefits.

One likely objection is fundamentally visceral: the charge that we are creating another bureaucracy. Such a creation presumably is bad a priori, and the objection is understandable. It will not, however, bear analysis. A bureau is merely a tool of social organization. As such, it must be evaluated in terms of its output rather than its mere existence. Clearly the incentive structures in bureaucracies often lead to socially costly outcomes associated with goal displacement, growth past the point where marginal social costs exceed marginal social benefits, and a host of other pathologies. In this case, however, we harness this incentive structure as a negative feedback to counter the pathologies. Analogies to this situation are common in medical biochemistry.

The second objection may be that the BBC may kill some worthwhile programs. Indeed it might. All drugs, especially the most useful, do kill some patients. So do seat belts. Is the agency, however, beneficial on net? Clearly such an agency would prey on the programs that are the most vulnerable to attack; that is, those whose social payoffs are demonstrably negative. The size of the BBC is, to put it crudely, a function of the stupidity of the prey agencies. A series of successful attacks is very likely to have a profound effect on the learning curve of the various agencies. Successful attacks are likely to generate doubts regarding the worth of other programs. Since the agencies are uncertain about which of their programs may be subject to predation and since all are fair game, they will have strong incentives to avoid proposing projects of dubious social utility. Should this be the case, policy is likely to be more carefully analyzed. The implications for the economics profession are obvious.

When writing in this area, it is increasingly difficult to end cheerfully. Our proposal is merely the first attack on a difficult problem. We realize that it needs refinement. While our proposal may seem radical when first considered, it will seem less so if reconsidered. In view of the direction the United States has been heading, there will be increasing incentives to reconsider. Given the avowedly experimental character of our political institutions, some proposal similar to ours is likely to capture the attention of at least one political entrepreneur. It is, in the final analysis, difficult to constrain a fundamental optimism regarding the ability of people to learn when self-interest is at stake.

Notes

1. Ralph K. Winter, Jr., "The Welfare State and the Decline of Electoral Politics," *Regulation,* March-April 1978, pp. 11–14.

2. Public goods are those goods for which (*a*) consumption by one individual does not interfere with the consumption of the same good by another individual, and (*b*) any individual cannot be excluded by price from enjoying the benefits of the goods' provision, once the good is produced. See Paul A. Samuelson, "The Pure Theory of Public Expenditures," *Review of Economics and Statistics* 36 (November 1954):387–89.

3. A. C. Pigou, *The Economics of Welfare,* 4th ed. (New York: Macmillan Co., 1932), esp. chap. 9.

4. The example is from John A. Baden and Richard L. Stroup, "The Environmental Costs of Government Action," *Policy Review,* Spring 1978, pp. 28–29.

5. Terry L. Anderson and P. J. Hill, *The Birth of a Transfer Society* (Stanford, Calif.: Hoover Institution Press, 1980), p. 5. Used by permission.

6. To request that individuals behave counter to their interests violates the cardinal rule of public policy: "Never ask a person to act against his own self-interest." See Garrett Hardin, *The Limits of Altruism* (Bloomington: Indiana University Press, 1977), p. 27.

7. See Garrett Hardin, "The Tragedy of the Commons," *Science* 162 (1968):1243–48.

8. Francis E. Rourke, *Bureaucracy, Politics, and Public Policy* (New York: Little, Brown, 1976).

9. Richard B. McKenzie and Gordon Tullock, *The New World of Economics* (Homewood, Ill.: Richard D. Irwin, 1975), pp. 196–210.

10. Herbert Simon, "Theories of Decision-Making in Economics and Behavioral Science," *American Economic Review,* June 1959. Reprinted in E. Mansfield, ed., *Microeconomics: Selected Readings,* 2d ed. (New York: W. W. Norton, 1975).

11. George W. Ladd, "Utility Maximization Sufficient for Long-Run Survival," *Journal of Political Economy,* July-August 1968, pp. 478–83.

12. Hardin, "Tragedy of the Commons."

13. Gardiner Ackley, "The Costs of Inflation," *American Economic Review* 68 (May 1978):149–58.

14. Stephen Miller, "A Good Word for Bureaucracy," *The American Spectator* 12, no. 3 (February 1979):14.

15. Ibid., p. 16.

Property Rights as a Common Pool Resource

by Terry L. Anderson and Peter J. Hill

Since environmental degradation can be attributed to the lack of well-defined and well-enforced property rights, it has been commonplace to turn to the government for solutions to these problems. These solutions can be categorized either as those that attempt to define and enforce property rights to resources or those that attempt to directly allocate resources. Literature on the environmental crisis is replete with studies that show the impact of the solutions on efficiency. In essence, these studies have attempted to measure the traditional welfare triangle, which economists use as an indicator of the magnitude of inefficiency.

To be sure, estimates of this nature are important, but this inefficiency is by no means the only loss. As studies of the welfare loss due to monopolies have shown, the size of the welfare triangle is small compared to the inefficiency that has come from "rent seeking." In the case of monopoly, this rent seeking arises from firms attempting to gain the rents that accrue to the monopoly. Since these attempts use valuable resources and produce nothing, inefficiency in addition to the welfare triangle occurs. To the extent that the establishment of property rights and the allocation of resources generate similar rents to the recipients of the rights or allocation, we can expect rent seeking to occur in these areas as well. This paper attempts to show how governmental policy has promoted inefficiency through rent seeking regardless of its impact on the welfare triangle. In other words, even though efforts to get the public domain into private hands have re-

This article was completed while Terry Anderson was a National Fellow at the Hoover Institution, 1977–78. For an expanded discussion of this issue, see Terry L. Anderson and Peter J. Hill, *The Birth of a Transfer Society* (Stanford, Calif.: Hoover Institution Press, 1980).

duced inefficiency by equating private and social costs and benefits, the manner in which private property rights have been established has wasted resources through the rent seeking process.

Our argument is couched in terms of the American frontier, not because it is specific to the frontier, but because the frontier provides an opportunity to compare alternative methods of defining and enforcing private rights. Over a half century ago, Frederick Jackson Turner published *The Frontier in American History*, which contained two basic themes. First, American institutions were shaped by the changing resource constraints that the pioneers faced as they ventured into the geographic fringes of their society.

> From the first, it became evident that these men had means of supplementing their individual activity by informal combinations. One of the things that impressed all early travelers in the United States was the capacity for extra-legal, voluntary association. This was natural enough; in all America we can study the process by which in a new land social customs form and crystalize into law.[1]

Second, the frontier provided a safety valve for labor that placed a lower limit on the wages that would be paid. After documenting the population increase in the Mississippi Valley relative to the Atlantic coast, Turner states

> These figures show the significance of the Mississippi Valley in its pressure upon the older section by the competition of its cheap lands, its abundant harvests, and its drainage of the labor supply. All of these things meant an upward lift to the Eastern wage earner.[2]

Our intention is to cast Turner's hypotheses in terms of an economic theory of property rights. In a general way, these hypotheses explain the evolution of property rights in response to changing relative prices of resources. This suggests that as more information is obtained about resources and as their value rises, potential rents to the resources will induce decision makers to utilize them and to define and enforce property rights that govern them. The nature of these property rights, or rules of the game, will reflect relative factor scarcity. Moreover, when the methods of defining and enforcing these rights are devised through collective user action, there is a greater incentive to conserve resources used in the process than when that process is imposed exogenously, even though both may lead to ownership institutions that guide resources to their highest valued use. Finally, we theorize that as the frontier closes (as the number of unclaimed resources declines), the safety valve will no longer exist, and individual options for increasing wealth will be reduced to productive activity or transfer activity or both.

Economics of the Frontier

The frontier is an area where natural resources are abundant relative to labor and where institutions governing the use of those resources are not yet developed. As long as the value of the marginal product of labor combined with those resources is less than the opportunity cost of labor in nonfrontier regions, the frontier resources will not be exploited. However, as the demand for output produced from natural resources increases or as the opportunity cost of resources falls, labor inputs will move to the frontier.

Following figure 1, the first unit of labor to enter the frontier will equate the wage rate with the value of his marginal product and devote L_1 units of labor to resource exploitation, let us say farming. He will capture a rent to the resource equal to area $ABCD$. With nonexclusive rights to the resource and without collusion among potential laborers, "rent becomes a residual, with every decision-making unit . . . maximizing the portion left behind by others."[3] Cheung has shown that when institutions governing the use of resources are not developed, all of the rents to resources will be dissipated through increased farming effort. In the limit, L_2 total units of labor will be commited to farming. Economic waste results, since the marginal product of all effort is less than the opportunity costs from alternative occupations, and may even be negative.[4]

Looking at the problem from another view, if fewer units of labor were devoted to farming, social output would rise by the amount of rents to natural resources. In his example of the fishery, Cheung puts it this way:

> there exist incentives to fishermen to restrict the *number* of decision units who have access to the fishing right. That is, even if each decision unit is free to commit the amount of fishing effort, the "rent" captured by each will be larger the smaller the number of decision units.[5]

The point is that potential rents to resources are valuable and, therefore, will attract efforts to capture them. One capture method is through increased fishing effort; another is through the definition and enforcement of exclusive rights to use the resource. As we have pointed out elsewhere, "establishing and protecting property rights is very much a productive activity toward which resources can be devoted."[6] The amount of effort devoted to this activity is a function of many variables,[7] but the important point is that as the potential rents from resources rise with the value of the resource, *ceteris paribus,* more efforts will be devoted to definition and enforcement activity. Individuals would be willing to spend up to the net present value of the rents to obtain perfectly defined and enforced rights. If the rents can be obtained for less, net rents will be positive and society's output will be greater. But even if all rents are dissipated through definition and enforcement, the fact that rights are established will reduce inefficiency by making decision makers responsible for their actions.[8]

Whether or not rents will be dissipated by definition and enforcement activity, however, depends on the process whereby rights

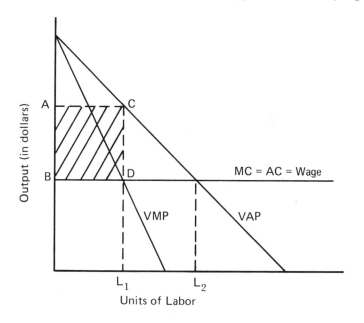

Fig. 1. Dissipation of rents under common ownership. *VMP* = value of marginal product; *VAP* = value of average product; *MC* = marginal cost; *AC* = average cost.

evolve. If two individuals must divide an acre of land between them, they have an incentive to accomplish the division in the cheapest manner, since they are responsible for the resources used in the bargaining process. Assume that the rent to the acre would be $100 if private rights were well defined. If each individual spends ten hours valued at $1 per hour in the bargaining process and the land is divided in half, individual wealth for each will rise by $40, i.e., total rent less resources used in bargaining. Reducing the number of hours spent to five will increase net wealth for each to $45.

As long as those bargaining for the property rights are free to choose their own definition process, there is an incentive to reduce bargaining costs. This is not to say that they will always resolve the issue and define property rights at a low cost. The prisoner dilemma problem, in fact, may even preclude any agreement. Indeed, since definition will affect the distribution of wealth, views of equity and justice will play a role in the final division. Indian battles with whites can be seen in this context. Since the two parties had very different ideas about equity and since each believed it had a right to the land, a low-cost division could not be reached. Furthermore, if it is technically difficult to measure the resource in question, definition may be very costly. Rights to scenery, air, and water provide examples. These problems notwithstanding, there is still an incentive to conserve on definition and enforcement resources when mechanisms are chosen by the participants.

The incentive for reducing bargaining costs will not be as strong if the agent establishing the rules, e.g., a government official, has no claim to the residual.[9] Returning to the land division example, suppose a third party chooses the decision process. Allocating the property to whichever individual arrives at the location first might seem fair, but many resources can be wasted in the process. Assuming that participants could not rig the race by agreeing that one person would walk to the property while the other waited at the starting line and that they would later split the property fifty-fifty, many inputs could be invested in the race.[10] Without collusion, each would be willing to spend up to the expected value of the rent to win the race. With probabilities of success accurately estimated, all potential rents would be dissipated. As the number of entrants to the race increases, the likelihood of collusion declines and the potential for rent dissipation rises.

At this point it is not possible to predict whether a definition and enforcement process established by user bargaining or government rules will result in less rent dissipation.[11] As the number of potential claimants increases, the cost of agreeing on distribution rises. Furthermore, as exemplified by conflicts between settlers and Indians, a third party with coercive power may be necessary to reduce waste in the definition and enforcement process. The process of defining and enforcing property rights will consume some resources regardless of its origin. The question is which process will consume the fewest resources. We can say that, when the process derives from those who will claim the residual, there is a built-in incentive to conserve resources, and there is evidence that this incentive produces the desired results.

We have discussed how the lure of unclaimed rents induces individuals to attempt to increase their wealth by defining and enforcing private rights; but how can wealth be increased when there are no unclaimed resources? The standard economics textbook answer is that improved efficiency is the only legal means. In other words, with all property privately held, individuals can increase wealth by moving resources from lower to higher valued alternatives or by altering the amount of output obtained from a given amount of inputs. There is nothing wrong with this answer, but it is incomplete. It may also be possible for an individual or group to obtain the private rights of others through nonmarket means. Theft, of course, is one way of doing this, but it is not the only way. The coercive power of government is often used. If some resources are owned collectively and allocated by the government, or if the coercive power of government can be used to transfer privately owned resources between individual members of society, then individuals will invest inputs into transfer activity for the same reasons that they invest in definition and enforcement activity. Those who are successful will have higher wealth at the expense of someone else. The net result, however, is not simply zero sum. Nothing will be produced, but resources will be expended in attempting to obtain and prevent transfers. Resources used in this way represent social waste, so the game becomes negative sum.[12] Whether and under what conditions the institutional framework allows for transfers is a complex subject.[13] Suffice it to say here that by the late nineteenth century this avenue of individual wealth accumulation was available.

But how does this relate to the frontier and the Turner thesis? The safety valve hypothesis is commensurate in part with seeking rents to unclaimed resources.[14] By moving to the frontier and obtaining ownership of resources, an individual's wealth would include a

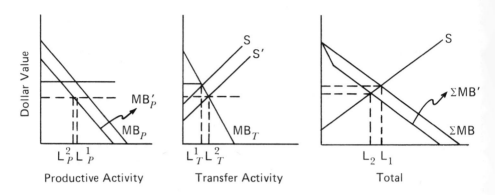

Fig. 2. Distribution of effort between productive and transfer activity. MB = marginal benefit; S = supply of effort for transfer activity.

return to labor *and* a return to resources. The closing of the frontier, however, eliminated the second possibility. As the opportunity for defining and enforcing rights to unclaimed resources declines, *ceteris paribus*, transfer activity becomes more attractive. This situation is shown in figure 2 by a shift to the left of the marginal benefit of productive activity (MB to MB'), thus increasing the residual supply of effort for transfer activity (S to S'). Total effort will decline to L_2; productive effort will decline to L^2_P, but transfer effort will rise to L^2_T.[15] To the extent that transfers are possible within the institutional framework, resources will be wasted, and the implications of the closing of the frontier can be significant.

We have argued that as individuals move to the frontier attracted by the relatively higher marginal productivity of labor combined with resources, they will also invest in the establishment of property rights over those resources. Whether or not this investment dissipates all of the rents to the resources will depend on the process by which property rights are defined and enforced. As long as the people determining this process have some claim to the residual, there is an incentive to conserve resources used in definition and enforcement activity. Once rights to resources are claimed, alternatives for increasing individual wealth are limited to improved efficiency or transfer activity. *Ceteris paribus*, we can expect a rise in efforts to use governmental authority to affect transfers, a process that has the characteristics of a negative sum game.

The Nineteenth-Century American Frontier

After the Revolution, the federal government received custody of all lands outside the boundaries of the thirteen original states and was charged with getting that land into the hands of private citizens. The development of a distribution policy was not easy, and it became even more complex as the country acquired resources through wars, trades, or purchases. For most of the nineteenth century, the policy that evolved had two dominant themes: (a) get the lands into private ownership as quickly as possible, and (b) promote the family farm, which Jefferson considered the backbone of our democracy.

To accomplish the distribution of public land, the resources were either awarded to squatters, sold, granted to promote the production of certain goods or services, or given to those willing to make certain investments in the land. The Ordinance of 1785 and the Land Laws of 1796, 1800, 1804, and 1820 governed different areas of the frontier and varied mostly in terms of the minimum size of land to be sold, the price, and the terms of payment.[16] The minimum price, which went from $1.00 to $2.00 and back to $1.25 per acre, appears to have been higher than that which would have derived from a simple auction. This and the rapid settlement of new territory encouraged squatting. As early as 1830, preemption of the squatters' rights was limited, but the Preemption Act of 1842 insured that squatters could buy their land at the minimum price. Finally, in 1862, the pressure toward giving land to the settlers culminated with the passage of the Homestead Act, followed by the Timber Culture Act in 1873, the Desert Land Act in 1877, and the Timber Stone Act in 1878. In all of these cases, however, there were strings attached to the "giveaway." First, the size of grants was limited below what was optimal. Second, the transfer into private hands was not complete until certain improvements had been made, including plowing, building, planting trees, and irrigating.

Through these and other acts a large portion of the arable land was transferred to private ownership, and much of the timber and mineral land was being logged and mined. But the late nineteenth century marked a dramatic change in federal disposal policy. The conservation movement, which was gaining momentum, equated "exploitation" with private enterprise and private ownership.[17] The result was a massive reservation of public lands and resources. By the twentieth century, the era of disposal of the public domain was over.

With this background, we will now apply the theory presented above to three phases of development in the American West, with the understanding that there is considerable overlap among them. During the first phase, extralegal voluntary associations preceded federal and state (territory) governments as individuals attempted to reduce rent dissipation by establishing private rights to resources. The methods used for defining and enforcing rights during this phase conserved on bargaining costs. The second phase saw the introduction of federal laws that encouraged wasteful use of resources in the definition and enforcement process. During the final phase, the United States entered an area of governmental transfer activity wherein increasing amounts of resources were used to obtain or protect ownership claims from other individuals.

Extralegal Voluntary Associations

Even before Turner published his frontier thesis, visitors to America commented on the propensity of the frontier to foster extralegal institutions for defining and enforcing property rights.[18] These institutions were efforts by frontiersmen, in the absence of formal government, to eliminate the rent dissipation that accompanies common ownership. The resources and assets covered by land clubs or claims associations, cattlemen's associations, mining camps, and wagon trains included land, water, livestock, minerals, timber, and personal property. The mechanisms used to establish private rights varied according to the asset in question, but, if the theory is correct, an effort should have been made in all cases to conserve the real resources consumed in the definition and enforcement process. Evidence from land claims clubs and livestock associations suggests that this was the case.

Claims associations were established to register settlers' claims to land and to insure that the claims would be honored when the federal government opened the land for private acquisition. Each club adopted its own constitution and bylaws, elected officers, established rules for adjudicating disputes, and most important, established the procedure for registering claims.[19] Where lands were surveyed, this entailed providing the clerk of the association with the range, township, and quarter section of the claim. Where there was no survey, stakes or blazes were to mark the boundaries, and the clerk was to be given a physical description of the claim, including landmarks and the names of surrounding claimants. If outsiders attempted to intrude upon a claim, the club would provide coercive power, if necessary, to enforce the member's rights. When the land was later put up for sale by the government, the association attempted to insure that the squatters could buy the land at the minimum price. In addition to providing private ownership for members, another alleged purpose was to protect the farmers from speculators, and much of the historical research on the clubs has debated whether they accomplished this task.[20] The question for our purpose, however, is not for whom they established rights, but what procedure was followed for defining and enforcing those rights.

The definition process followed the relatively low-cost method, and the available evidence on enforcement activity is consistent with the hypothesis that voluntary associations have an incentive to discourage rent dissipation. The requirements for improving claims varied. In two Iowa counties, Webster and Poweshiek, claimants had to expend labor worth $10 for each month after the first month, $30 during the first six months, and $30 for every succeeding six months.[21] Obviously such procedures, if enforced, did require resources.[22] In Johnson County, on the other hand, resident members were not required to invest any real resources in their land until they so desired. Nonresidents did have to expend $50 worth of labor for each six months the claim was held. While this may have encouraged speculation, it also conserved resources invested in enforcement and encouraged the optimal time path of investment in land. In his study of "The Iowa Claim Clubs," Allan Bogue concludes that "regulations prescribing the degree to which the member must improve his claim appear in the manuscript records of the Poweshiek, Johnson, and Webster county associations, but *not* in the selections of the club laws printed in the histories of other counties."[23] Since there were between 25 and 100 claims associations in Iowa at this time,[24] it is reasonable to conclude that these voluntary organizations did not encourage the dissipation of rents through enforcement activity.

Cattlemen and livestock associations throughout the West also worked to define and enforce property rights and to conserve resources devoted to these activities. In *The Cattlemen's Frontier*, Louis Pelzer describes the place of these associations in the formation of "frontier law."

> From successive frontiers of our American history have developed needed customs, laws, and organizations. The era of fur-trading produced its hunters, its barter, and the great fur companies; on the mining frontier came the staked claims and the vigilance committees; the camp meeting and the circuit rider

were heard on the religious outposts; on the margins of settlement the claim clubs protected the rights of the squatter farmers; on the ranchmen's frontier the millions of cattle, the vast ranges, the ranches, and the cattle companies produced pools and local, district, territorial, and national cattle associations.[25]

The basic purpose of these associations was to restrict entry into the open ranges. As Granville Stuart wrote, "The business was a fascinating one and profitable so long as the ranges were not overstocked."[26] As long as the grass was free and use was unrestricted, entry into the business would continue until all rents were dissipated. Only with restricted entry could these rents be maintained.[27]

Two methods were used for parceling out the range and preventing entry.[28] The first was "squatter sovereignty," or prior use, which was as old as the frontier itself; the second depended on the acquisition of water rights. During the early settlement of the Great Plains the first method was sufficient. "There was room enough for all, and when a cattleman rode up some likely valley or across some well-grassed divide and found cattle thereon, he looked elsewhere for range."[29] The following newspaper advertisement attests to the ease with which cattlemen were able to define their "range rights."

> I, the undersigned, do hereby notify the public that I claim the valley, branching off the Glendive Creek, four miles east of Allard, and extending to its source on the South side of the Northern Pacific Railroad as a stock range.—Chas. S. Johnson[30]

As Osgood says, however, such a claim was not enforceable in any court of law; it could only be enforced through the local cattlemen's associations.

Such groups also used local newspapers to announce their intentions to enforce the "accustomed range."

> We the undersigned, stock growers of the above described range, hereby give notice that we consider said range already overstocked; therefore we positively decline allowing any outside parties or any parties locating herds upon this range the use of our corrals, nor will they be permitted to join us in any roundup on said range from and after this date.[31]

Since individual ranchers were hard pressed to survive without "participation in the roundup, in the use of the common corrals, in the group protection against Indians, thieves, and predatory animals and, in some cases, in the group drive of the beef turnoff to the railroad,"[32] the enforcement procedure was effective.

The alternative to livestock associations was to obtain ownership of water rights. In some cases, this was accomplished by purchasing from squatters the land adjacent to the water, and in others it was accomplished through homesteading. Where state or territorial laws allowed filing for water rights, such action was taken, and the range would extend from the stream to the divide separating it from the next drainage. This doctrine of "prior appropriation," which eventually became a distinguishing feature of western water law, enabled the cattleman to establish his private rights to the range.[33]

As our theory suggests, both the claims clubs and the cattlemen's associations organized to establish rights over scarce resources and, in so doing, provided a definition and enforcement process that conserved resources expended in these activities. In some claims clubs, land improvements were required to insure ownership; but in most, filing a claim was sufficient.

The case of definition and enforcement of range rights is less clear. Resources were certainly expended to prevent rent dissipation through common ownership. That the stockmen accepted newspaper claim advertisements and tied establishment efforts to other ranch activities, however, suggests that, as residual claimants of the rents, they did attempt to conserve on definition and enforcement activity. To be sure, they did not require the construction of unnecessary buildings or the utilization of inappropriate ranching techniques. Such requirements would have consumed valuable resources— resources for which the ranchers had alternative uses.

Law from the East

> The Easterner, with his background of forest and farm, could not always understand the man of the cattle kingdom. One went on foot, the other went on horseback; one carried his law in books, the other carried it strapped round his waist. One represented tradition, the other represented innovation; one responded to convention, the other responded to necessity and evolved his own conventions. Yet the man of the timber and the town made the law for the man of the plain; the plainsman, finding this law unsuited to his needs, broke it and was called lawless.[34]

The federal requirements for distribution of the public domain never seemed to suit the frontiersmen. From the Ordinance of 1785 to the Homestead Act of 1862, complaints ranged from prices that were too high to size limitations that were too low. The Easterner did have a different frame of reference and a different set of goals. Perhaps more important for our analysis, the lawmakers who set the rules for definition and enforcement of property rights did not have the incentive to reduce wasteful rent dissipation.[35]

There are ways in which the public domain might have been transferred into private hands while conserving resources devoted to property rights establishment. Ignoring the revenue problem for the government, public resources could have been auctioned off to the highest bidder. The virtue of this system is that it requires that buyers only exchange money or claims to use resources rather than expending the resources themselves. Taylor Dennen effectively explains this process.

> Under the auction system, settlers will exchange money, that is, a command over resources, for a land title. However, with the price of land set at zero, individuals will use up real resources to get to the land at the moment when its present value turns from negative to zero. . . . The problem is fully analogous to the dissipation of consumer's surplus under price controls of rationing by waiting. That is, where prices are inhibited in their role of allocating commodities, consumers will use up valuable resources such as capital equipment or labor time in order to obtain commodities. From the point of view of those interested in maximizing the value of output, such resource expenditure is pure waste.[36]

Some of the early land laws resembled this system; they disposed of the land by auction with the provision that a minimum price be paid. As long as the actual present value of the land exceeded the minimum price, no resources were wasted. But the minimum price did exceed the present value and, therefore, encouraged squatting.

To the extent that claims associations succeeded in establishing private rights without resource waste and to the extent that preemption rights were limited, the disposal policy during the first half of the nineteenth century was relatively efficient. Two other aspects of federal land policy also encouraged efficiency. First, the Graduation Act of 1854 provided for successive reductions in the minimum price at which unsold public lands would be offered. Though the act was only in effect for eight years, it did approach the more simple auction described above. Second, millions of acres of public domain were granted to private enterprise, mainly the transcontinental railroads, and they in turn sold it to the highest bidders. While the equity considerations of this system may be questioned, it did discourage rent dissipation through definition and enforcement activity.

The shift in public domain disposition policy that accompanied the Homestead Act of 1862 reserved some of these efficiency gains. This and similar acts explicitly required an expenditure of labor or capital or both as a condition for ownership. The quantity of resources wasted by this process is difficult to estimate; some of the investments would have been made without the law, some would have been postponed, and some would never have been made. Libecap and Johnson estimate that one timber company spent between $660 and $870 for entryman payments, cabins, and agents in order to establish rights to 160-acre plots of California timber land.[37] Stories abound of resources used in land rushes.

> For example, considerable time was spent simply waiting, or jock-eying for an advantageous position at the starting line. On occasion special vehicles were constructed which would presumably speed more quickly over the land to a claim site. Supposedly in one instance an individual who had a right to enter the unsettled territory spent considerable time training a pony to race to a particularly good claim.[38]

On the other hand, "ingenuity" (commonly called fraud) probably worked to mitigate some of these resource costs. Cabins sworn to be 12 by 14 but measured in inches rather than feet and structures on wheels that could be moved from one claim to another were used as proof of residence. Single plow furrows filled with water were used as proof of irrigation. Moreover, the fact that "the great bulk of the land put into cultivation after 1865 was purchased from federal and state government and from land-grant railroads"[39] served to mitigate rent dissipation through real resource expenditures. The extent to which the government's method of public domain disposal wasted resources must await further research, but there is evidence that it did not work as efficiently as it might have. Twenty-seven percent of the over one billion acres disposed of passed through the homestead system, and 285 million acres were taken up more than once. From this Dennen concludes, "There is no reason to believe that the impact of the federal land-disposition system on the national economy was insignificant."[40]

The Rise of Transfer Activity

The last quarter of the nineteenth century witnessed not only a significant change in policy regarding disposal of the public domain, but also a fundamental change in the sanctity of individual liberty and private property rights. In 1923, John Burgess described our political and constitutional history as "an almost unbroken march in the direction of a more and more perfect individual liberty and immunity against the powers of government, and a more and more complete and efficient organization and operation of sovereignty back of both government and liberty, limiting the powers of government defining and guaranteeing individual liberty."[41] This about-face, however, left "hardly an individual immunity against government power which may not be set aside by government, at its own will and discretion, with or without reason, as government itself may determine."[42] The turnabout is explained in part by postbellum economic conditions:

> The gains from post–Civil War economic growth are taken for granted, whereas the associated costs are emphasized and the business leaders of the day are cited with shame as "robber barons." Economic expansion and per capita income growth brought about a sweeping transformation in the structure of the economy and engendered disruption at the same time. Business fluctuations were sharp and frequent and generated high rates of unemployment. Innovations reduced costs but cast aside those whose businesses or skills tied them to outmoded processes. Urbanization produced slums and pestilence as well as factory unemployment. Growth in government came as a by-product and brought with it a host of problems in public administration and policy-making.[43]

A full explanation of the reduced confidence in a laissez faire organization of the society and the increased intrusion on individual rights is very complex and beyond the scope of this paper.[44] The increasing willingness and ability of the government to tamper with private rights coupled with the closing of the frontier, however, do fit into the rent-seeking framework and have serious implications for resource allocation.

The ability of the government to interfere with the allocation of privately owned resources was a prerequisite for the growth of transfer activity that blossomed in the late nineteenth century. Earlier Supreme Court interpretation of the Constitution, especially under Chief Justice John Marshall, made it clear that the government was not to impair the obligation of contracts, interfere with interstate commerce, or in any way violate the vested or natural rights of citizens. The due process clauses of both the Fifth and Fourteenth Amendments were initially applied to criminal cases and used to insure that specific procedures were followed in prosecution. No deprivation of life, liberty, or property was considered "reasonable." Moreover, profits earned through the private enterprise system were emphasized and considered good for the individual as well as the country. "Pursuit of profit not only justified doing ordinary competitive damage to a market rival, but was a business firm's whole legal excuse for being."[45] According to a Michigan court, "a business corporation is organized and carried on primarily for the profit of stockholders. ... It is not within the lawful powers of a board of directors to shape and conduct the affairs of a corporation for the merely incidental benefit of shareholders and for the primary purpose of benefiting others."[46] Such an institutional environment led to what James Willard Hurst has called "the release of energy."[47]

While it is not our intention to discuss in detail the many constitutional changes that occurred during the postbellum period, certain Supreme Court cases were significant. The era witnessed a "revolution in the due process of law,"[48] with the Supreme Court adding substance to the Fourteenth Amendment. Three important cases illustrate these changes.

First, the Slaughterhouse Cases (1873) opened the door for substantive due process or "reasonableness." The plaintiffs challenged a Louisiana law granting a monopoly to a New Orleans slaughterhouse on the grounds that it violated their rights to due process. The majority of the court dismissed the claim, but in the dissent, Justice Noah H. Swayne defined due process as the "*fair* and *regular* course of procedure."[49] The stage was set for later decisions that would determine the reasonableness of state regulation of individual rights.

The second case, *Munn* v. *Illinois* (1877), made reasonable regulation a reality. The constitutionality of regulating grain elevator storage rates was the issue. Chief Justice Waite concluded in the majority opinion that when "one devotes his property to a use in which the public has an interest, he, in effect, grants to the public an interest in that use, and must submit to be controlled by the public for the common good, to the extent of the interest he has thus created."[50] Justice Field recognized the pathbreaking nature of this decision. In his dissent, he said, "If this be sound law, if there be no protection, either in the principles upon which our republican government is founded, or in the prohibitions of the Constitution against such invasion of private rights, all property and all business in the State are held at the mercy of a majority of its legislature."[51] The idea that private property used in the public interest was subject to public regulation gave the government a broad new authority to interfere with contracts and private property.

Finally, *Muller* v. *Oregon* (1908) cemented the rule of reasonableness in American constitutional law. The issue of the regulation of working hours for women was important for labor legislation, but the manner in which the case was argued was paramount. Louis Brandeis was retained by the State of Oregon to argue its side and, in support of the "reasonableness" of the statute, he filed a brief consisting of only two pages dealing with constitutional arguments and over a hundred pages dealing with social conditions.

The importance of these cases and the general institutional environment that prevailed is that they allowed the use of the coercive power of government for the transfer of rights from one individual or group to another. Since individual wealth is in large part a direct function of rights held, the effect was to raise the marginal benefit of employing resources in transfer activity. To the extent that existing rights are available to anyone, they are analogous to common pool resources and the same problems of rent dissipation occur. The result is the social waste of resources. Moreover, the policy of not allowing private ownership of many resources meant that such resources would be allocated by the government rather than by markets. To the extent that governmental allocation does not fully utilize prices, and for the most part it does not, individuals will employ resources to establish rights to what is being allocated, thereby dissipating rents.

The closing of the frontier compounded the impact of the increased return on transfer activity brought about by the institutional environment. As pointed out above, the frontier allowed individuals the opportunity to increase their wealth by establishing rights to unclaimed resources. The closing of that possibility, *ceteris paribus,* increases the supply of resources to transfer activity (see figure 2). There can be little doubt that the frontier was closing. The rise in average population density from 10.6 persons per square mile in 1860 to 35.6 in 1920 tells much of the story. By the turn of the century, most of what was left to be claimed was less productive or less accessible. The possibility of establishing individual property rights over previously unclaimed natural resources diminished with each passing year and each new homestead. Lawrence M. Friedman, in his *History of American Law,* puts it this way:

> What really passed was not the frontier, but the idea of the frontier. This inner sense of change was one of the most important influences on American law. . . . By 1900, if one can speak about so slippery a thing as dominant public opinion, that opinion saw a narrowing sky, a dead frontier, life as a struggle for position, competition as a zero-sum game, the economy as a pie to be divided, not a ladder stretching out beyond the horizon.[52]

Transfer activity had found its place in American society.

There are two indications that the extent of transfer activity was not trivial. The first and most direct measure is an upper-bound estimate of the resources timber companies might be willing to expend to get federal stumpage. Table 1 shows the commercial volume, government receipts from sales, estimated market value of timber cut on national forests, and rent for the period 1910 to 1955. For the

TABLE 1. National Forest Timber Volume, Receipts, and Value

Year	Volume[a] 1	Government Receipts[b] 2	Market Value[c] 3	Rent Col. 3 − Col. 2 4
1910	379	$906	$1,099	$193
1915	547	1,165	1,477	312
1920	783	1,748	2,153	405
1925	1,005	2,793	2,864	71
1930	1,470	4,324	5,071	747
1935	649	1,701	1,330	−371
1940	1,347	3,803	3,031	−772
1945	2,712	11,663	14,374	2,711
1950	3,195	29,084	55,433	26,349
1955	6,225	70,105	171,187	101,082

a. In millions of board feet

b. In thousands of dollars

c. Computed using the average price of Douglas fir and ponderosa pine. *Historical Statistics of the U.S.* (Washington, D.C.: Government Printing Office, 1960), pp. 316–17. The value is even higher if southern and sugar pine prices are included, but most national forest stumpage would be in Douglas fir and ponderosa pine. See Marion Clawson, *The Federal Lands* (Baltimore: Johns Hopkins University Press, 1957), pp. 406–7.

years listed, the Forest Service sales price was below the market stumpage price for all but two observations. Assuming that private wealth maximizers would compete to dissipate the rents (the difference between actual government sales price and market value), they should be willing to expend up to the amount of the rents. Since the government allowed the timber cut on national forests to be sold below the market clearing price, there would have been an excess demand in the absence of any alternative rationing processes. Assuming that Forest Service officials were not bribed, it is likely that companies lobbied, "wined and dined," and used other efforts to obtain contracts. The point is not to accuse officials of corruption or companies of illegal practices, but to show that when prices are not used to ration goods, other mechanisms will be used that can consume valuable resources. If all of the rents were dissipated through the employment of valuable resources, the potential waste in the allocation of timber alone would have been in the millions of dollars. Since other natural resources, such as mineral and grazing rights, recreation, and water, are rationed in a manner similar to timber, wasteful rent dissipation can be expected in these areas as well. We have argued that individuals or groups attempting to increase their wealth through transfer activity engage in a negative sum game for society. In the case of governmental allocation of resources through nonprice processes, the magnitude of this sum appears to be quite large.

The second indication of transfer activity is the rise of associations and groups whose purpose is to obtain or prevent governmental transfers. This indicator is less precise, but it potentially consumes a larger volume of resources. A review of the progressive movement of the late nineteenth and early twentieth centuries is useful here. Many historians have characterized the progressive and the conservation movements as attempts to gain public control of private wealth. Rose M. Stahl suggests that the struggle against corporate power had two phases:

> first and most important, was the attempt to find some adequate means of controlling and regulating the corporate activities; and the second, and almost as important, was the resistance to the efforts of corporations to exploit the natural resources of the nation in their own behalf.[53]

The Supreme Court decisions discussed above marked the success of the first phase and the reservation of the public domain the success of the second. The data in table 1 illustrate the magnitude of transfer activity that results when the government allocates resources, but they tell us nothing of the transfer activity inherent in the regulation. This regulation involved the transfer of decision-making rights and property rights from one individual or group to another. Often the issue was over input and output pricing, but it also included explicit control of privately owned inputs and outputs. One can attempt to justify this type of regulation as an effort to improve market imperfections or to eliminate externalities, but this is not our point. Again, we wish to stress that transfer activity entails the employment of scarce resources in a potentially negative sum game.

The rise of interest groups and organizations during the last quarter of the nineteenth and the early years of the twentieth century reflects our entry into a transfer society. Lawrence Friedman describes the impact of these groups.

> organization in the last half of the 19th century was more than a matter of clubs and societies. Noticeably, many strong interest groups developed—labor unions, industrial combines, farmers' organizations, occupational associations—to jockey for position and power in society. These groups molded, dominated, shaped American law.
>
> A group or association has two aspects: it defines some persons in, and some persons out. People joined together in groups not simply for mutual help, but to exclude, to define an enemy, to make common cause against outsiders. Organization was a law of life, not merely because life was so complicated, but also because life seemed so much a zero-sum game.[54]

Transfers themselves are zero sum since they take from one and give to another, but this taking and giving does not come about by itself. The groups themselves were organized to affect or prevent transfers, depending on whether they were giving or receiving. John C. Calhoun said, "Power can only be resisted by power—and tendency by tendency"[55]—and groups provided that power. "Men and women joined in large, staffed, and disciplined program groups such as our public life had not seen before."[56] Examples include the Grange (1867), the Farmers' Alliance (1874, 1887), the Knights of Labor (1869), the American Federation of Labor (1881), the American Women Suffrage Association and the National Woman Suffrage Association (1869), the United States Brewers Association (1862), the National Potters Association (1875), the Stove Founders National Defense Association (1886), and the National Association of Manufacturers (1895). Those specifically dealing with natural resource policy include the National Rivers and Harbors Congress, the American National Livestock Association and state livestock associations, the National Water Users Association, the American Forestry Association, the American Association for the Advancement of Science, the Audubon Society, the American Game Protective Association, the Boone Crockett Club, the National Geographic Society, and the Sierra Club. In 1974, there were no less than 264 international, national, and interstate organizations listed in the *Conservation Directory*.[57] To be sure, not all of the efforts of every organization go to transfer activity, but for many a significant proportion of their endeavors are directed at determining the rules that establish who has access to resources, both physical and human, and the outputs therefrom.

Two disclaimers are in order. First, we recognize that not all members were interested in transfer activity and not all group activity was aimed at transfer activity. Some undoubtedly joined simply for comradeship, and some joined because the organization *produced* real goods and services. Second, we do not claim that the organizations were wasting their members' contributions or that they might not have had some positive impact on social output. From the member's viewpoint, the groups were certainly producing a valuable service; membership was voluntary. Whether they increased social output has been and will continue to be debated.

Samuel P. Hays states, "The deepest significance of the conservation movement,... lies in its political implications: how should resource decisions be made and by whom? Each resource problem involved conflicts. Should they be resolved through partisan politics, through compromise among competing groups, or through judicial decision?"[58] Resolving these conflicts through a political process that requires the investment of scarce resources rather than through a pricing process that only involves the exchange of rights to resources consumes scarce inputs. When all that is produced is a transfer, there is social waste.

Notes

1. Frederick Jackson Turner, *The Frontier in American History* (New York: Krieger, 1920), p. 343.

2. Ibid., p. 193.

3. See, for example, Fred A. Shannon, "The Homestead Act and the Labor Surplus," *American Historical Review* 41 (1936):637–51.

4. Steven N. S. Cheung, "The Structure of a Contract and the Theory of a Non-Exclusive Resource," *Journal of Law and Economics* 13 (1970):50.

5. Ibid., p. 54.

6. Terry L. Anderson and P. J. Hill, "The Evolution of Property Rights: A Study of the American West," *Journal of Law and Economics* 18 (April 1975): 165.

7. For a more complete discussion, see Terry L. Anderson and P. J. Hill, "Toward a General Theory of Institutional Change," in *Frontiers of Economics* (Blacksburg, Va.: University Publications, 1976), pp. 3–18.

8. For a discussion of the optimal time path for investment in resources, see R. Taylor Dennen, "Some Efficiency Effects of Nineteenth-Century Federal Land Policy: A Dynamic Analysis," *Agricultural History* 51 (October 1977).

9. It is beyond the scope of this paper to establish the incentives facing governmental officials who establish the process. Nevertheless, it is an important issue. To explain why they chose homesteading instead of an auction requires specification of their objective function. Assuming that bribes and corruption were not commonplace, we can be certain that rent maximization was not their goal.

10. For a brief discussion of this type of investment, see Dennen, "Some Efficiency Effects."

11. Because of prisoner dilemma problems, it is convenient to believe that externally imposed solutions will waste fewer resources. That position, however, runs counter to the public choice school of thought that focuses on incentives, or reasons for believing that certain solutions will result. Externally imposed property rights assignments often hold the *possibility* of resource savings. The real question is how *probable* is it that such resource savings will result.

12. We use the terms negative zero and positive sum with reference to the production of goods and services of the society. Of course, it is possible for a game that is negative sum in goods and services to be positive sum in utility terms. Because of the difficulty (impossibility?) of making interpersonal utility comparisons, we choose not to cast our discussion in that framework.

13. For a more complete discussion, see Terry L. Anderson and P. J. Hill, *The Birth of a Transfer Society* (Stanford, Calif.: Hoover Institution Press, 1980); idem, "The Independent Judiciary in a Rent-Seeking Perspective" (Bozeman: Department of Economics, Montana State University, 1978).

14. The safety valve thesis may also suggest that the marginal product of labor combined with frontier resources is higher.

15. Some additional effort may be devoted to productive activity. The distribution between transfer activity and productivity will depend on the elasticity of demand in the two areas.

16. See Lance E. Davis et al., *American Economic Growth: An Economist's History of the United States* (New York: Harper & Row, 1972), pp. 104–5.

17. See Samuel P. Hays, *Conservation and the Gospel of Efficiency: The Progressive Conservation Movement, 1890–1920* (Cambridge: Harvard University Press, 1959), chap. 13.

18. See, for example, Alexis de Tocqueville, *Democracy in America,* ed. J. P. Mayer and Max Lerner, 2 vols. (New York: Harper & Row, 1966), pp. 283, 492–96.

19. For an example of a constitution of a claims club, see Benjamin F. Shambaugh, "Frontier Land Clubs, or Claim Associations," *Annual Report of the American Historical Association* (1900).

20. See Allan G. Bogue, "The Iowa Claim Clubs: Symbol and Substance," in *The Public Lands: Studies in the History of the Public Domain,* ed. Vernon Carstensen (Madison: University of Wisconsin Press, 1963).

21. Ibid., p. 51.

22. Presumably, the resources were productive in that they improved the land in some way. The question is whether that investment would have been made anyway and if the expenditures were less than those required by the government under the Homestead Act.

23. Bogue, "Iowa Claim Clubs," p. 51. Emphasis added.

24. Shambaugh, in "Frontier Land Clubs," says there were 100 associations in Iowa, while Bogue says there were only 25.

25. Louis Pelzer, *The Cattlemen's Frontier* (Glendale, Calif.: Russell Sage, 1936), p. 87.

26. Granville Stuart, *Forty Years on the Frontier,* ed. Paul C. Phillips, vol. 2 (Cleveland: Arthur H. Clark Co., 1925), p. 185.

27. For a detailed discussion of the theory of common property as it applies to the open range problem in the West, see R. Taylor Dennen, "Cattlemen's Associations and Property Rights in Land in the American West," *Explorations in Economic History* 13 (October 1976).

28. Ibid., p. 426.

29. Ernest Staples Osgood, *The Day of the Cattleman* (Chicago: University of Chicago Press, 1929), p. 182.

30. Quoted in ibid., p. 183.

31. Quoted in Dennen, "Cattlemen's Associations," p. 427.

32. Osgood, *The Day of the Cattleman,* p. 185.

33. For a more detailed discussion of water rights in the American West, see Anderson and Hill, "Evolution of Property Rights."

34. Walter Prescott Webb, *The Great Plains* (Boston: Grosset and Dunlap, 1931), p. 206.

35. For specific discussions of the Homestead Acts, see Dennen, "Some Efficiency Effects"; and Robert Higgs, *The Transformation of the American Economy, 1865–1914* (New York: John Wiley & Sons, 1971).

36. Dennen, "Some Efficiency Effects," pp. 729–30.

37. Gary D. Libecap and Ronald N. Johnson, "Property Rights, Nineteenth Century Federal Timber Policy, and the Conservation Movement," mimeographed (Albuquerque: Department of Economics, University of New Mexico, 1978).

38. Dennen, "Some Efficiency Effects," p. 730.

39. Higgs, *Transformation,* p. 91.

40. Dennen, "Some Efficiency Effects," p. 736.

41. John W. Burgess, *Recent Changes in American Constitutional Theory* (New York: Arno Press, 1923), p. 1.

42. Ibid.

43. Ralph Gray and John M. Peterson, *Economic Development of the United States,* rev. ed. (Homewood, Ill.: Richard D. Irwin, 1974), p. 269.

44. For a more detailed discussion, see Anderson and Hill, *Birth of a Transfer Society.*

45. James Willard Hurst, *Laws and the Conditions of Freedom in the Nineteenth-Century United States* (Madison: University of Wisconsin Press, 1964), p. 76.

46. Quoted in ibid.

47. Ibid., chap. 1.

48. Alfred H. Kelly and Winfred A. Harbison, *The American Constitution,* 4th ed. (New York: W. W. Norton, 1970), chap. 19.

49. Ibid., p. 510. Emphasis added.

50. 94 U.S. 126 (1877).

51. 94 U.S. 140 (1877).

52. Lawrence M. Friedman, *A History of American Law* (New York: Simon and Schuster, 1973), p. 296.

53. Rose M. Stahl, "The Ballinger-Pinchot Controversy," *Smith College Studies in History* 11 (January 1926):69.

54. Friedman, *A History of American Law,* pp. 296–97.

55. Quoted in Hurst, *Laws and the Conditions of Freedom,* p. 42.

56. Ibid., p. 87.

57. William E. Clark, ed., *Conservation Directory* (Washington, D.C.: National Wildlife Federation, 1974).

58. Hays, *Conservation and the Gospel of Efficiency,* p. 271.

Social Costs, Incentive Structures, and Environmental Policies

by Lloyd D. Orr

Environmental pollution represents a classic case of the failure of private markets to allocate resources effectively when transaction costs are high and property rights are poorly defined. Collective intervention is necessary to avoid the destructive effects of treating natural resources in the public domain as though they were free goods.

Unfortunately, collective intervention is not synonymous with good policy. There are deficiencies in present policy that have serious short-run and long-run consequences. A strong case has been made that our environmental policies (*a*) inherently create excessive costs relative to their impact, (*b*) destroy private property values unnecessarily and arbitrarily, and (*c*) inhibit the operation of the decentralized decision structure that is essential to establishing and maintaining a high quality environment.

This article will not yield a simple, comprehensive solution to problems of environmental quality. It will give weight to the view held by most economists that effluent charges are a vastly underutilized basis for sound environmental policy. My suggestion is that effluent charge strategy can be recast to promise less and deliver more—thus becoming economically, administratively, and even politically more viable.

Appreciation is due to Del Gardner and Vincent Ostrom for helpful comments on an earlier draft. This article originally appeared in *Western Political Quarterly* 32, no. 3 (September 1979):286–97. Reprinted by permission of the University of Utah, copyright holder.

Basic Principles

Certain basic technical and economic factors should surround any critique of environmental policy. These principles enable us to focus on the fundamental characteristics of the pollution problem and confine alternative solutions to those that are in accord with physical and economic realities. A brief discussion of these principles includes conservation of energy and mass, common property resources and market failure, the role of technology, and the business firm as an intermediary.

Conservation

Neither energy nor matter is destroyed, but they are both readily altered from one form to another in nature and by human activities. This is an application of the first and second laws of thermodynamics. An intuitive grasp of these laws is necessary if we are to understand the fundamentals of environmental and resource conservation.[1]

The assertion that our economy produces goods and services for our use establishes an image of a product that is created (such as an automobile) and then gradually destroyed by use over a period of years. The image is misleading. Both production and consumption are transformation processes. Resources are transformed from their natural state to a state that man finds more useful. Iron ore is transformed to steel, crude oil into gasoline, and so forth. The process continues as products are used or wear out, which is to say they are transformed into yet other states, such as rusty metal and carbon dioxide. The first law of thermodynamics, which is a strict law of conservation, tells us that whatever goes into the production process as a natural resource ultimately emerges as a residual placed back into the environment.

Many of these residuals force themselves into our consciousness and receive the label "pollution." The process may take many years, as with a house or an automobile, or it may take place quickly, as in the pumping, refining, and burning of a petroleum product. The important point is that it takes place as an *inevitable* result of economic activity. The process represents a movement of material from concentrated or usable natural resources to dissipated and less usable residuals. This is an application of the second law of thermodynamics, which deals with the concept of entropy.

The following is a simple illustration of these principles as applied to the production and consumption process.

Natural resources, *A,* are used to fabricate products, *B,* with the resulting residual outputs, *C.* Use of the products in turn results in their transformation to residuals, *D.* Strict conservation requires:

$A = B + C$:   Natural resources emerge from the production sector as products or industrial residuals.

$B = D$:   Products are used up and become residuals of the consuming sector.

Therefore,

$A = D + C$:   Natural resources are ultimately transformed into production or consumption residuals.

Whatever goes in from the environment ultimately emerges as a residual.

An initial environmentalist reaction may be that the outlook is bleak, since it appears that the only way to reduce residuals is to reduce output, a solution suggested by some. Recognition of thermodynamic implications shows us that some pollution control merely transforms one kind of residual into another. If the second residual is less environmentally harmful than the first, and the resource inputs (e.g., sewage treatment plants) required are not too great, even this may be a good thing. However, there are many other things that we can do. The trick is to increase the *ratio* of *services* from products (B) to total residuals ($C + D$).

An excellent example of these services (which we shall call B^1) is automobile travel. We want to reduce the amount of residuals created by a given amount of automobile travel and thus increase the ratio $B^1/C + D$. The primary ways to accomplish this are as follows:

1. To increase the energy and material efficiency of the product.
2. To efficiently recycle materials so that they become residuals at a much slower pace.
3. To increase the durability of the product.

In addition, we may shape the *form* and geographic placement of the residuals so that they make better use of the natural assimilative capacity of the environment and are less harmful to us.

Clearly, we cannot stop the creation of residuals, but we may be able to effectively reduce and maintain them at levels where a healthy environment and a high standard of living are compatible. It is a matter of the age-old economic question of *choice*.

Market Failures

For various interrelated reasons of history, ubiquity, and contracting costs, environmental resources such as atmosphere and waterways are not held privately and are not marketed or subjected to cost calculations. Their common property status dictates that their conservation is dependent on *public choice* and some form of public management. If we have a "pollution problem," it is because of a reluctance to make such choices or to establish effective machinery for implementation.[2]

The combustion process provides one illustration of the importance of property rights and markets. Both the automotive engineer and the driver pay a great deal of attention to the quantity of gasoline that passes through the carburetor, engine, and exhaust system of an automobile. Traditionally, neither paid any attention to the quantity and quality of atmosphere that passes through the same system, even though atmosphere is just as important as gasoline for the combustion process. The reason is simple. The gasoline is controlled by a private or public body and has an explicit cost. As a common property resource, the atmosphere is available to all of us with no explicit cost. Like any resource that is treated as though it were free, it gets "overused," and we all suffer the implicit costs of pollution as third party effects of private contracts between automobile owners and oil companies.

Because a large number of people are involved in the use of a ubiquitous environment, the costs of private contracts for its use are prohibitively high. The ultimate result is that the private costs of marketed goods and services are too low, while the total social costs—which include environmental degradation—are too high. The existence of common property resources and the resulting third party effects of private contracts create inefficiencies in the economic system.

There are some third party effects associated with almost any private contract, but they are usually trivial, and the costs of overcoming them are high enough that they should be ignored—perhaps more often than they are. Although there are large differences of opinion, few people would place environmental problems in this category. They are serious problems and are certain to become more so as socioeconomic systems increase in density and complexity. Not only are questions of efficiency involved, but questions of equity arise in the differences among people who receive the private goods and those who bear the costs of pollution. The fact that pollution control policies will inevitably involve some sacrifice of human freedom cannot be raised as an automatic barrier to collective action. The very existence of serious third party effects indicates that individuals are coerced to bear costs of private contracts in which they do not participate. Clearly, we must concern ourselves with the *forms* of freedom as well as with efficiency.

Ultimately, we must choose at the margin among broad social goals, such as freedom, equity, and efficiency, in exactly the same way we must choose among more tangible goods. None of these goals is absolute. We always have and always will make choices among them. But innovation in public policy is analogous to technical innovation in private production. It can make more of all social goods available.

Technology

The role of technical innovation in private production needs to be briefly reviewed because its relationship to pollution is so badly misunderstood. Technology is often condemned as the cause of our environmental and natural resource problems, but technology is neutral. If technology goes astray, it is because society gives the wrong signals to its managers about what it wants to conserve. If the signal is that environmental resources are free, we should not be surprised to find a technical response that leads to their profligate use. Conversely, if an *appropriately structured* signal is given (and reinforced by incentives) that these resources are scarce (costly) and are likely to become more costly, those same technological forces will be unleashed in the relatively unexplored area of environmental conservation.

Pollution and Profits

The role of the business firm and the business community in the creation of pollution is also poorly understood. The caricature is that capitalist greed causes pollution and that "high profits" come at the expense of the environment. The nonsequitors are fairly obvious. The profit motive provides the incentive for production in a market economy. If environmental resources are underpriced they will be overused, and pollution will be excessive. But the capitalist is not the primary beneficiary in this process. Market analyses demonstrate that the major long-run result will be lower cost products for consumers, not higher profits for capitalists. *The firm is primarily an intermediary in the process of social choice between lower cost products and higher environmen-*

tal quality. This is generally true for both perfect and imperfect competition.[3] It is also true if firms are organized on a socialist rather than on a private basis. It is the citizen as consumer who has received the bulk of private benefits from environmental exploitation. It is the citizen as consumer and taxpayer who in the long run will pay the costs of environmental conservation. The problem is how society wants to allocate its resources. It is not basically a question of capitalist (or noncapitalist) greed.

If these claims are true, it is fair to ask why the business community is so resistant to environmental programs. The answer is clearly in the short-run problems of *transition,* from the view of commonly held property as a free good to the view that it is a scarce and expensive resource. This transition imposes costly discontinuities and uncertainties on individual firms and the industrial structure. It destroys returns to capital and other resources in some areas and enhances them in others. These costs are initially felt and resisted by the business community. They are ultimately borne by society through the destruction of some of its capital and human resource values.

The basic tenets of good policy are obvious, and they are the same from both the private and public points of view: (*a*) whatever degree of environmental quality we seek, we should try to buy it at minimum cost over the long run; and (*b*) low transition costs—*ease of transition*—is a vital element of successful policy. As we shall see, our present policy is poorly conceived with respect to these criteria.

Regulation of Water Quality

In this section, the description of pollution control policy will be confined to water quality regulation as administered by the Environmental Protection Agency (EPA) and individual states.[4] The general approach and results are similar for other areas of environmental quality regulation and, in fact, bear a discouraging generic relationship to many areas of government regulation.

The responsibility for water quality control has increasingly shifted from the states to the federal government over the past thirty years. Within the federal government, programs administered by diverse agencies have become more concentrated. In 1970, the EPA was created and now administers the major federal programs.

Federal water pollution control policy is based on the Water Pollution Control Act of 1948 as amended on several occasions and culminating in the major amendments of 1972. The amendments represent attempts to deal with the continual decline in water quality in the face of federal, state, and local efforts to reverse the trend. The resulting federal-state plan embodies three important features.

First, the scope of the plan is comprehensive in the coverage of all navigable and all interstate waterways. States must apply for and receive EPA approval of a program that meets national goals. If a state fails to apply for or receive approval, the EPA administers a program for the waterways within the state. These requirements were a response to the states' failure to implement or enforce effective water quality programs.

Second, after effluent standards are established for all industries that discharge water-borne wastes, they are enforced through a national permit system. State programs are essentially programs for the administration of the national permit system. This technique of point source regulation is intended to circumvent the legal and administrative difficulties of previous programs, where restrictions on polluters had to be tied to their responsibility for the failure of a body of water to meet water quality standards. The basis of current regulation is not the water quality itself, but the permitted point source discharge. The basis for discharge permits is that "best practicable technology" (BPT) be installed by July 1, 1977, and that "best available technology" (BAT) be installed by July 1, 1983. The average of an industry group provides the basis for BPT. The best performance in an industry provides the basis for "economically achievable" BAT. Fines for violations range from $2,500 to $25,000 a day for the first offense and up to $50,000 a day for subsequent offenses. Congress has set a goal of zero pollutant discharge into navigable waterways by 1985.

Third, publicly owned treatment facilities are handled separately under the 1972 amendments. Standards require secondary treatment for all sewage plants. Federal subsidies are available for up to 75 percent of construction costs on municipal plants. Applicants must demonstrate that the planned project is sound, needed, and consistent with federal water pollution control policy. Each recipient of waste treatment services from a subsidized plant must pay a proportionate share of the total costs for such services. This requirement is the basis for the growing incidence of surcharges for extra strength industrial effluent. The primary direct subsidy available to industry is the accelerated depreciation on capital expenditures made for industrial waste treatment facilities that was contained in the Tax Reform Act of 1969. Further subsidy is available in the expanded use of tax-exempt bond financing for industrial pollution control facilities.

Essentially, U.S. policy involves court enforcement of administered point source standards based on legislative directive. It is an attempt to curtail an undesirable activity in a situation where the effects of the activity and its control are pervasive and complex. Substantial authority is passed to a regulatory agency that is insulated from responsibility for *the most important consequences* of its decisions. This effect is enhanced when major costs, both direct and indirect, take place in the private sector where they are largely concealed from those who ultimately pay them—consumers.

It should not be surprising that efficiency incentives are not only lacking, they are actually perverse. Environmental policy is an example of the inability of the political process to move beyond a simple-minded concept of police powers toward the required restructuring of incentives.[5] Pollution is not the result of antisocial behavior by a few individuals. It is the result of actions by everyone in the pursuit of socially desirable goals. The purpose of the policy is to induce people to continue their desirable activities, but in ways that reduce the amount of pollution produced. Police powers and criminal sanctions, with their "all or nothing" characteristics, are poor instruments for accomplishing this purpose. This reliance furthers excessive interference with the desirable activity and generates costs that are disproportionately higher than derived benefits. The major impacts discussed below were predictable and indeed were predicted on the basis of political-economic analysis and experience with pre–1972 legislation.[6]

Social Costs of Environmental Policy

The basic indictment of water quality regulations and environmental policy generally is that they are socially wasteful. Resource misallocation is the result of the poor structuring of incentives in a complex situation that requires a heavy reliance on incentives for success. In addition, dependence on administrative law leads inevitably to a substantive denial of due process.

Kneese and Schultze estimate the total cost of the 1972 water quality amendments to be at $220 billion by the early 1980s.[7] When they add nonpoint source costs for water pollution control and the costs for the similarly structured air quality regulations, they suggest a "conservative estimate" of $500 billion in pollution control outlays alone. Outlays of $50–60 billion per year are estimated for the early 1980s. The authors necessarily stress the crude and uncertain nature of these estimates because they do not include economic dislocation and many other indirect costs.

The estimates represent substantial amounts of resources, even for an economy that is expected to produce between two and three trillion dollars in output per year during the early 1980s. But these estimates fail to explain the social waste of present policies, and they would provide only limited insight if the total benefits expected from the programs were known.[8] We can learn more from looking at some of the major distortions built into present programs and the warped incentive structures with which we will face the even greater environmental quality problems of the 1980s. Whatever difficulties we now face, they are likely to worsen as the environmental bureaucracy ages and as public attention is diverted to other issues and crises.

Efficiency in meeting a water quality standard dictates that a given amount of pollutant be removed at minimum cost. Efforts should be concentrated where removal is the easiest. Current policy dictates *uniform* treatment for both industrial and municipal point sources. It directly frustrates the efficiency goal and wastes resources that could be used to buy more environmental quality (pollution control equipment, research and development, leisure) or other economic goods desired by the community.

Subsidies provided directly to municipalities and industries through the tax laws create several distortions in the process of pollution control. Only capital expenditures for treatment plants are subsidized. This biases pollution control in favor of "end of pipe treatment" at the expense of many possible process changes that would be adopted as more efficient if the polluter had to pay the full cost. Here are three results of such policies. First, process changes within industrial plants that would reduce the generation of effluent are seen as too costly since they are *not* generally subsidized. Second, segregation of higher quality waste by industry and municipalities would enhance reclamation. Water may be reclaimed for various industrial and public uses and sludge may be more safely processed into high quality fertilizer. These processes are often inhibited by uniform standards and subsidization of a concept that aggregates all sewage at the central plant. Third, in the case of municipalities, the subsidy has been observed to create a bias in favor of capital at the expense of maintenance and operating outlays. A GAO study of the pre–1972 subsidy program concludes that over half of the plants surveyed were operated inefficiently due to insufficient operating outlays.[9] Subsidy also creates incentives for municipalities to delay making needed repairs and improvements. If federal largess is based on "need," the obvious strategy is to do as little as possible in order to maximize the federal contribution toward "needed" facilities.

How environmental policy creates a time-sequenced adaptation to the environment as a scarce resource is much more crucial than the issues raised above, although not independent of them. This process involves problems of transition, responses to transitional burdens, and the institutional structures that we are building for the future. These are crucial issues because they determine future potential for efficiency and desirable social structure, as well as the current effectiveness of policies.

The very concept of issuing permits to over 60,000 point sources of water pollution with targets for future years based on best practicable and best available technology stirs notions of administrative nightmares. Following complex guidelines, the effective administrator must consider engineering aspects of control technology, process changes, environmental and conservation side effects, and "such other factors as the Administrator deems appropriate." Economic costs and benefits are to be considered, although the language is different in the two cases of BPT and BAT. There is broad latitude in these guidelines, and it would be virtually impossible to provide specific language that dealt realistically with each situation.

The language of the law has clearly established the EPA as a centralized agency charged with the detailed control of all point source water pollution. Therefore, the proliferation of EPA guidelines is subject to legal challenge, and the dominant incentives for private and public polluters are to ignore, delay, challenge, and fight. In the bargaining situation that often results, the community often sides with the threatened polluter. Political considerations dominate the outcome, while benefit-cost ratios tend to get lost.

Once the technical and legal process has been completed with a particular firm or municipality, the situation becomes static. Management deals with the next item on its agenda, and the EPA moves on to the next standard-setting and permit-issuing problem. The firm has no established incentive to husband environmental resources. The installation of BPT may well impede the pursuit of BAT due to the durability of capital installations, the possibility of marked differences in the technologies involved in the two standards, and the "economic considerations" in EPA guidelines. Innovation is stymied in still other ways. Consider the incentive of a manager to be innovative in pollution control when his success may then become the standard for BAT in all of his plants, according to a timetable determined by an outside agency.

These large and intractable administrative problems seem small only when compared with what will happen to them with the growth and technological change anticipated for the 1980s and beyond. The basic difficulty is an absence of effective institution building. The lack of incentive to seek low-cost solutions now and to adapt technologically on a continuous and decentralized basis in the future is a serious shortcoming of current policy. *The transition to a society where each decision maker regards the environment as a valuable resource never really takes place.* We can predict cycles of costly response to environmental crises.

An important side effect of the bargaining situation and finite EPA funding is selective enforcement. Those facing and actually complying with the toughest standards are likely to be those least able to resist, for a variety of political and economic reasons. The difficulty for the small firm is that it may not have the resources either to fight or to comply. Such firms are usually referred to as "old and obsolete" in bureaucratic, if not economic, jargon. The uncertainties inherent in the standard setting and enforcement process and the limited ability of these firms to absorb sudden overhead costs make them vulnerable to these arbitrary regulations, while their demise may or may not reduce pollution. Moreover, the consideration of alternatives to standard setting is limited by law.

It would be difficult to measure the effect of environmental regulation on industrial structure, but the direction is clearly toward survival of the biggest. Much of the current difficulty experienced by Chrysler and American Motors can be traced to a combination of environmental, safety, and fuel efficiency regulations. This is sobering, since these are giant corporations by almost any but automobile industry standards.

A more direct example involves a 1977 congressional mandate on the pretreatment of wastes to be released into public sewer systems. Prior to December 1979, the EPA was required to "promulgate specific and detailed standards setting forth what treatment equipment must be installed by 21 industrial categories and scores of subcategories." The industries must install the technology no later than December 1982. Small electroplating shops are expected to be especially hard hit with the closing of more than 580 shops and the loss of 12,000 jobs.[10] The contribution of these small shops to the toxic waste problem within a cost-benefit framework has not, to my knowledge, been specified.

A paraphrase from Kneese and Schultze neatly summarizes the problem. The regulatory approach suffers from an inescapable dilemma. If the approach is simple enough to be handled by a centralized agency, it is bound to be inefficient. If it seeks to accommodate the diversity and complexity of the economy in devising efficient effluent standards, the regulatory task becomes insurmountable. The unavoidable arbitrariness and inequity will make court enforcement difficult, time consuming, and ultimately ineffective.

Transition and Innovation: A Reappraisal of Effluent Charges

The purpose of environmental policy is to attain a reasonable allocation of society's environmental resources.[11] Most economists favor effluent charges or some similar market substitute as a basis for sound policy. These charges are generally viewed as an important and vastly underutilized policy tool rather than as a solution for all environmental problems.

The economic analysis of effluent charges has centered on static equilibrium and welfare principles. Effluent charges are regarded as analogous to market prices. The selection of the proper effluent charge will force the polluter to face a marginal cost that is equal to the marginal damage caused by his discharge. The result is an optimal level of pollution control at minimum cost.

This analysis promises much more than any policy can deliver. We cannot measure the marginal damage from pollution, even for major pollutants, with anything close to the precision required for setting optimal charges. We also lack the information required to establish the charge levels needed to achieve *any* preselected level of pollution control. Following Baumol and Oates, economists have moved to the position of advocating effluent charges as a means of meeting politically determined environmental standards at minimum cost. The proposed solution establishes charge levels through an iterative (trial and error) procedure that gradually approaches the charge structure required to meet the predetermined standards. The procedure is essentially learning by doing.

There are, however, many practical problems associated with the implementation of these charge structures. There are monitoring difficulties, problems of geographical differences in charge levels, and the likelihood of excessive adjustment costs associated with a trial and error search for appropriate charges. Anthony Dorcey notes that the iterative process could lead to the very sort of protracted bargaining and regulatory problems from which effluent charges are supposed to provide an escape.

The difficulties faced by the standards-effluent charge approach are basically the same as for the traditional regulatory approach. It is extremely difficult to establish a national program with enough flexibility to meet the diverse and complex requirements for pollution abatement.[12] There is a substantial advantage in the decentralized incentive structure that is inherent in effluent charges, but much of this advantage may be dissipated by uncertainties and disincentives that are created by the iterative process. A response to one charge level may be made obsolete by a subsequent change. The resulting costs would be borne by both owners and customers. Attenuated response and resistance to change can be anticipated. The management problem begins to look very much like the one we have with present policies.

An escape from this dilemma involves the realization that, given our limited regulatory abilities, we are trying to regulate in the wrong domain. We attempt a series of complex, short-run, and discrete adaptations to a problem with basic characteristics that are long run and continuous. The policy goal should be a smooth transition to a set of decentralized decision-making institutions that continuously respond technologically to the environment as a scarce resource. Rather than trying to match the complexity of the problem directly, we should try to build self-regulating institutions with the requisite complexity.

The importance of fostering technological change is always stated in discussions of effluent charge strategies, but it has only recently begun to receive emphasis.[13] What is generally missing is the view that the *fundamental* strategy of environmental policy should be to provide a framework for *continuous* and detailed technological adaptation to the environmental impacts of growth, change in location, and change in product and process technology. This shifts the emphasis from meeting standards to growth with scarce and exhaustible resources.

Economic historians provide us with valuable insight on the long-run importance of technical change as a response to resource scarcities.[14] In their investigations of technical progress as a response to relative factor scarcities and the related question of factor "bias" in innovation, the analytical distinction between technical change and the static concept of resource substitution became extremely difficult to maintain. In Rosenberg's words, "*Today's* [resource] substitution possibilities are made possible by *yesterday's* technological innovations." We thus see the problem of ease in substitution as essentially one of innovation in creating substitution possibilities, and the problem of relief from future environmental resource scarcities comes to rest squarely on our ability to induce technological response.

Historians also indirectly provide us with the best evidence in support of effluent charges as a technique for inducing the required innovation. In their assessment of the importance of technological change in past growth, they provide valuable documentation (*a*) on the vast range of technological response possibilities to particular resource scarcities and (*b*) on the *detailed adaptation to such scarcities over time under the spur of decentralized cost incentives.*[15]

There are a variety of reports on direct experience with effluent charges, usually in connection with municipal sewer charges and in-plant water and waste charges.[16] The data are sparse, and the results need to be interpreted with caution, but there are clear illustrations of the cost incentive effects that are inherent in effluent charges. Managers respond to resource cost in ways that are suited to their own particular circumstances. The nature of the response would be difficult to predict, either for someone trying to plan a specific result or even for the waste dischargers themselves. The reports strengthen the notion that environment-saving technology is a sparsely explored area that yields quickly to proper incentives. As a result, charges have been effective, their effectiveness has increased with time, and they apparently have not been excessively disruptive. In many cases, innovation has a favorable effect on cost in addition to the reduced effluent charges.

Recognition of the importance of innovation with its emphasis on the need for thorough, detailed, and continuous adaptation to the social costs of using common property resources highlights the most compelling advantage of effluent charges in comparison with alternative environmental policies. It also makes clear the extent of our ignorance about the appropriate nature of specific technical solutions. The administrative ability to dictate effective solutions to environmental problems at reasonable cost is extremely limited.

Our inability to "fine tune" environmental quality should be obvious regardless of the policy approach, but awareness of the limits of our knowledge should not lead to paralysis or the abandonment of environmental policy to traditional styles of regulation. Rather, it suggests gradualism as a means of minimizing the uncertainties of transition cost. It also suggests flexibility (decentralization) in response mechanisms to induce effective use of current and prospective knowledge and to guard against massive changes in a complex milieu where side effects are largely unknown. Effluent charges become very imperfect tools, but, in my view, they are clearly superior to known alternatives as a *basic* policy.

Emphasis on establishing a long-run framework of technological adaptation effectively reduces some of the economic and political problems associated with the implementation of effluent charge strategies. For example, many of the economic and political problems of effluent charges are related to transition costs. Gradual implementation in predetermined steps over a period of years should provide the needed inducement for innovation while minimizing short-term disruptions. Rather than being faced with highly uncertain levels of effluent charges (or regulatory standards), the waste discharger responds to a stable profile of increasing charges extending several years into the future. Environmental quality standards may still be relevant, but the *direction* and *rate of change* in environmental quality are criteria that should be of more continuing importance. *Time* and *appropriate incentives* are our best allies in reducing transition burdens.

As the above implies, there can be a marked reduction in the current stress on cumbersome and potentially costly iterative adjustments in charge structures. Proper charge levels remain an important issue, and we need more information to establish a desirable *ordering* of charges for the major pollutants at levels that will induce an initial response. The experiences we have noted suggest that this may be surprisingly easy. The iterative process would still be useful where initial charge structures fail to elicit an adequate response or where the burdens are clearly onerous and not justified by an actual or anticipated response. Errors will still need to be corrected, but the abandonment of ambitions to fine-tune carries with it a marked decrease in the use of an iterative process.

Another implication of emphasis on innovative response is a decrease in the importance of regional differences in charge levels. Under effluent charge incentives, innovation in one region will tend to be adapted in cost sensitive ways to other regions. The possibility that regional differences could seriously affect the competitive position of plants *within* an industry at different locations is likely to make substantial differentials politically unacceptable. Also, since there is no way to predetermine the technological response to a basic nation-wide charge level, we cannot know that the gain in allocative efficiency from differential charges, based on current parameters, would be worth the costs of disruption and relocation. Thus, for both political and economic reasons it seems wise to start with gradual implementation of nationwide charges. If the short-run transition burdens are still too heavy, other business taxes could be reduced. This would provide general relief without destroying the incentive to reduce effluent or the valuable differentiation *between* industries on the basis of quantity and quality of waste discharge. In much the same way, redistributing revenue in ways that are not related to effluent discharge could provide general relief to municipalities that are hard pressed by effluent charges.

There may be a need for supplementary charges or regulations to deal with special problems at the regional or local level. While federal policy establishes a charge for each pound of sulfur emitted from a smokestack, a regional or local authority may determine the height of the smokestack according to local pollution problems. This is clearly appropriate, since the problems can be more narrowly specified at the subnational level. The ability to establish an effective regulatory response with sensitivity to the full range of consequences is thereby enhanced.

It is, of course, possible to conceptualize both perfect regulation and perfect charge structures. The problem is that we do not know enough and will not know enough in the foreseeable future to implement either type of plan. We need to choose wisely among recognizably *imperfect* plans.

Ignorance is not a good thing, but *recognizing* ignorance can help us avoid serious errors.[17] The strength of effluent charges lies in their ability to move things in generally desirable directions even when we lack the knowledge to provide charge structures that are efficient. The diversity of impact, continuous incentive properties, and a relatively high degree of self-regulation serve to limit the undesirable (and unforeseen) side effects and the perverse incentives that are associated with many regulatory policies. The relevant concept is *robustness,* the ability to function effectively in the face of the often serious differences between the ideal of the model and the reality of operative policy. This concept is relevant to all policy formulation, and its serious consideration should prevent some of the disaffection with the government that follows from recurrent policy failures.

Notes

1. For a more complete discussion, see John Dales, *Pollution, Property, and Prices* (Toronto: University of Toronto Press, 1968).

2. For a broad discussion of these issues, see Garrett Hardin and John Baden, *Managing the Commons* (San Francisco: Freeman, 1977).

3. For a statement to be strictly correct, all industries must be perfectly competitive and at constant cost. The most likely cause of loss to capital would be the possible destruction of monopoly rents (profits). Scherer roughly estimates these to cause a redistribution of income of 3 percent of GNP. This implies that, even in the ludicrously extreme case where "pricing" common property resources eliminated all monopoly profits, the redistribution of income from monopolists to others would be only 3 percent of GNP. See Warren Scherer, *Industrial Market Structure and Economic Performance* (Chicago: Rand McNally & Co., 1971), p. 409.

4. A fuller description is given in A. Kneese and C. Schultze, *Pollution, Prices and Public Policy* (Washington, D.C.: The Brookings Institution, 1975).

5. See Edwin Mills, *The Economics of Environmental Quality* (New York: W. W. Norton, 1978), pp. 202-5.

6. See A. M. Freeman III and R. H. Haveman, "Clean Rhetoric and Dirty Water," *Public Interest* 28 (Summer 1972):51-65; A. Kneese and B. Bower, *Managing Water Quality: Economics, Technology, Institutions* (Baltimore: Johns Hopkins University Press, 1968); C. Schultze, *The Public Use of Private Interest* (Washington, D.C.: The Brookings Institution, 1977); and Kneese and Schultze, *Pollution.*

7. Kneese and Schultze, *Pollution,* chap. 6.

8. Measurement of waste would require knowledge of marginal costs and benefits. It would also require a comparison of actual costs with the minimum necessary to achieve the task.

9. As reported by Freeman and Haveman, "Clean Rhetoric," p. 55.

10. New York Times News Service, July 28, 1978.

11. Parts of this section are drawn from L. Orr, "Incentive for Innovation as the Basis for Effluent Charge Strategy," *American Economic Review Proceedings* 66 (May 1976):441-47.

12. In cybernetics, the problem is stated as the "law of requisite variety." For complete control of an outcome, the regulator must have as much variety available to him in his regulatory machinery as there is in the disturbances presented to him in the area of regulation. The complexity of the economy and the environment faced by the regulator thus gives rise to the dilemma paraphrased above from Kneese and Schultze. Traditional regulatory concepts do not contain the requisite variety to efficiently control environmental pollution.

13. See Kneese and Schultze, *Pollution;* Orr, "Incentive for Innovation"; and Schultze, *Public Use of Private Interest.*

14. N. Rosenberg, *Technology and American Economic Growth* (New York: M. E. Sharpe, 1972); idem, "Innovative Response to Materials Shortages," *American Economic Review Proceedings* 63 (May 1973):111–18; and P. A. David, *Technical Choice, Innovation and Economic Growth* (New York: Cambridge University Press, 1975), pp. 57–91.

15. Rosenberg, "Innovative Response"; and G. Hueckel, "A Historical Approach to Future Economic Growth," *Science* 187 (March 1975):925-31.

16. J. A. Seagraves, "Industrial Waste Charges," *Journal of the Environmental Engineering Division, American Society of Civil Engineers* 99 (December 1973):875; Kneese and Bower, *Managing Water Quality*, pp. 168, 170-72; U.S., Congress, Joint Economic Committee, *Economic Analysis and Efficiency of Government, Part 6, Economic Incentives to Control Pollution: Hearings before the Subcommittee on Priorities and Economy in Government*, 91st Cong., 2d sess., March, 1971 (testimony of R. H. Havemann, L. Kimball, and J. Kinney).

17. Holling and Goldberg note the similarities between complex economic (urban) and ecological systems. Both appear to have considerable stability within certain bounds, but when the complex balance is disturbed by intervention, they can generate unexpected and undesirable results. Policy actions that are limited in scope and diverse in nature are suggested, using principles based more on recognition of ignorance than the presumption of knowledge about the nature of the system. See C. S. Holling and M. A. Goldberg, "Ecology and Planning," *American Institute of Planners Journal* 37 (July 1971):221-30.

U.S. Natural Gas Policy: An Autopsy

by Ernst R. Habicht, Jr.

It is a pleasure to dissect our late, lamented natural gas policy. Before I commence, however, I want to discuss an early leader in the oil business—that is, the whale oil business. Whale oil was produced jointly with an increasingly valued by-product, whalebone, which was much in demand for such things as foundation garments and buggy whips—items now obsolete as a result of either social or technological change. Similarly, natural gas, often produced jointly with oil, was initially demanded only in modest amounts and, when encountered on a significant scale associated with petroleum, was long considered a nuisance.[1] Gas has since gone through the typical maturation cycle of industrial commodities: curiosity, luxury, necessity, and, finally, national security.

Like its successors, the oil and gas industries, the whaling industry enjoyed its share of tax and regulatory incentives and constraints.[2] As early as 1645 in Southampton, Long Island, a bounty of five shillings was paid to anyone who located a stranded whale, *unless* the animal was discovered on the Sabbath (whereupon no bounty was paid).[3] Since no data are available, we can only speculate whether the incidence of Saturday and Monday whale discoveries exceeded Sunday finds in regular fashion. This may represent the first regulatory incentive for converting one type of oil discovery into another, at least in North America.

Early on, the Plymouth and Massachusetts Bay colonies made provisions to equally divide the profits that were obtained from stranded whales between the colonial government, the town having jurisdiction, and the finder. In 1662, the town of Eastham, on Cape Cod, voted a share of stranded whales to the support of the minister. Such actions placed both government and God on the side of the oil business, but they may also have had the effect of prematurely encouraging offshore exploration or, at least, swift clandestine export.

By the early eighteenth century, shore-based boat whaling was flourishing. The discovery of more attractive oil resources, particularly the accidental capture of the first sperm whale in 1712, and numerous advances in whaling technology led to swift adoption of ship whaling. Gradually, the whaling ports of Massachusetts, especially New Bedford, came to dominate the industry. Not surprisingly, the success of the industry attracted political attention.

Various New York governors imposed taxes on whaling proceeds and set forth industry regulations designed to divert the whaling trade from Massachusetts and Connecticut to New York. On a somewhat larger scale, the British government required that colonial oil and bone be exported to English ports, where duties were imposed, while British whalers were granted handsome bounties. Most of these carrots and sticks disappeared before the Revolutionary War, only to be revived with a vengeance afterwards when British duties soared as Massachusetts attempted to bolster its shattered industry with substantial bounties. Thus, the foundations for encouraging and protecting domestic oil production were set down decades, even centuries, before the petroleum age.

The government did not always facilitate industry. In 1793, Congress required that American whaling vessels carry numerous documents, but this law was not rigorously enforced for some years. When the Long Island-based, three-masted bark *Monmouth* returned to New York on May 24, 1839, with rather slim proceeds valued at $13,000, the customs collector unexpectedly questioned her documents, found some missing, and slapped a duty of $4,500 on her cargo. There were predictable protests, perhaps best summed up by the editor of the Sag Harbor, Long Island, *Corrector,* Harry Hunt, who referred to the abrupt revival of the law as a "wreckless piece of insolence."[4]

Taxation, subsidies, and local regulations contributed little to the decline of the whaling industry compared to the pressures of competition and diminishing resources. After the halcyon decades of 1820–60, whaling never recovered from the introduction of petroleum lubricants and gas lights, innovations that were hastened by the high price of whale oil products. For some years afterwards, the demand for whalebone remained high, as did its price, but this proved an insufficient base for the industry (until new technology and the demand for food coalesced to threaten the end of a number of species after World War II).

A century later, hoops, whips, corset stays, and other articles once made of whalebone were either obsolete or made of plastic, mostly produced from natural gas. Today, plastic models of the grand old whaling ships—and even of whales themselves—are available everywhere. They convey, I submit, a sad impression of the reality of the whale. Similarly, one could argue that rather different (mathematical) models give us an insight of dubious value into the nature of our energy policy.

America's Hidden Energy Policy: Regulation and Taxation

Any governmental policy dealing with a commodity like energy must, regardless of the nature of government, be a *pricing* policy. Prices have incentive effects; that is, they determine the demand for and the supply of goods and services. High prices inhibit demand and encourage investments in more supplies or they coax forth innovative alternatives to traditional supplies.[5] Low prices have precisely the reverse effect.

Government can influence energy prices in a number of ways. For example, it can tax some or all forms of energy and thereby raise energy prices. Conversely, it can give tax breaks and other subsidies for producing energy and, thus, lower energy prices while encouraging energy production (or at least not discouraging it). Government can also regulate energy prices directly, as it has done over the last twenty-four years for natural gas sold in interstate commerce. It can establish performance standards and penalties or impose import quotas and tariffs directed toward a variety of goals, all of which show up as energy price changes. It has the power to directly regulate utilities and thus can directly control their prices (rates) and investment decisions. Finally, government has the authority to ration (allocate) energy.

Over the last decade, the federal government has increasingly regulated the oil industry's prices for both crude oil and refined products, allocated these substances, and overseen the industry's investment policies. This can leave us with the worst possible mismatch of both worlds, private advantage and regulated public utilities. One example is the decision to build an oil pipeline across Alaska but a gas pipeline from the North Slope through Canada. A common corridor to the Midwest would have been a far superior choice and was advocated during the trans-Alaskan pipeline litigation nine years ago. Once built across Alaska, however, a strong environmental and economic case can be advanced for diversion of the oil pipeline's deliveries into international markets. However, the associated gas may not be economically produced and delivered for many years to come. Likely it will be made available earlier, at a considerable loss to the economy. This wastefulness, compounded by other irrational acts, cannot possibly be construed by outsiders as part of a "sensible" energy policy.

Once these pervasive governmental powers are understood, can we still say that energy policymaking is a brand new endeavor? Of course not.

The nation already has an energy policy, the same one we've had for decades. It is premised on the increasingly untenable assumption that energy is both cheap and abundant and will only grow more so with the passage of time. This assumption—based on the actual experience of many years—came to determine the way we regulate and tax the energy industry. But the policy is no longer valid, the central flaw being that it presumes *both* abundance *and* decreasing costs.

Our present energy policy severely discourages economically sound, technically practical investments in improving the efficiency with which we use energy (energy conservation), while it simultaneously encourages investments in technologically questionable and economically wasteful energy supplies. It strongly favors *existing* energy forms and institutions: oil or natural gas from conventional reservoirs; central station, utility owned electric generation; and the movement of utilities into gas production. At the same time, it discriminates against novel but more economical energy technologies, such as improved efficiency of energy use, solar energy, cogeneration, or production of natural gas from unconventional reservoirs, which cannot easily be institutionally assimilated with those energy forms we depend on today because of inflexible subsidy patterns.

Prevailing tax codes and utility regulatory policies primarily reward investing in conventional energy supplies or their expensive, utility supplied substitutes, like synthetic natural gas, regardless of their costs. Tax policies of particular importance are the investment tax credit and accelerated depreciation provisions of the Federal Tax Code, which pay little heed to either the productivity of new capital investment or to the useful life of economic assets. (As one respected analyst points out, the tax laws may not merely entrench inefficiency; they may be a foremost engine of monopoly.)[6] Where gas or electric utilities are involved, it is common to price these energy forms on an average basis and use declining block rates, so that the more gas or electricity a customer uses, the lower the cost of additional consumption, regardless of the costs of meeting added demands. In this way, energy users *never* see the replacement costs of energy. No wonder portions of the oil and gas producing industry and feedstock and fuel users either in unregulated markets or under heavy curtailment are among the few who have really "discovered" energy conservation. They are also the few who confront the full opportunity costs of oil and gas.

The Carter administration's proposals for a new energy policy, like those of the Nixon and Ford administrations, were not set forth in the context of an existing policy, even though they implicitly acknowledged certain of the problems created by prevailing tax and regulatory constraints. In this regard, each successive administration has done more homework and has advanced increasingly sophisticated (but necessarily more intricate) energy policy proposals. This, of course, has led to more alienation of a widening array of interest groups and to predictable political stalemates.[7] The tar babies of regulation and taxation are much decried of late but, at least where energy is concerned, a solid majority seems comfortably stuck fast to them.

The Corpus Delecti

Many incomplete autopsies have been performed on the natural gas industry. They focused on the producers' end of the sytem and variously praised or damned wellhead pricing and investment controls. My own interest stems from an initial concern for the users' end, which is under state or local regulation nearly everywhere in the country. We all recognize that the two ends are tied together physically, but does what goes on at one end make good sense for the other? Are the two ends tied together economically? Increasingly, the answer is no.

Some years ago, the Environmental Defense Fund (EDF) joined forces with a consumer group[8] to protest the increasing manufacture of synthetic natural gas (SNG) from petroleum products. We focused on the fundamental insanity of selling gas that costs about $5.00 per thousand cubic feet (Mcf) to gas users at a rolled-in price of no more than $2.00 per Mcf. Along with liquified natural gas (LNG) imports, SNG represents the worst outcome of outdated regulatory and tax incentives applied to a public utility. While a cynic may define a utility as the only business enterprise that can sell at a loss and make up the difference in volume, what effect has this on the nation as a whole?

Any *added* demand for oil or oil-derived products must be met with *imported* petroleum. When a new SNG plant goes into operation, the naphtha it uses probably comes—directly or via displacement—from Saudi Arabia or another OPEC nation with a significant amount of spare oil producing capacity. Imagine, if you can, a pipeline connecting an oil reservoir in Saudi Arabia to water heaters and industrial furnaces in Indiana that are fitted with an ingenious economic valve that somehow lowers the apparent cost of fuel by 60 percent along the way!

This absurd fiction has become regulatory reality; indeed, there are already over a dozen such commercial SNG plants around the country (one of which was financed by an investor-owned utility using tax free industrial revenue and pollution abatement bonds). More SNG plants are proposed each year, and their construction and operation have been actively encouraged by our present energy policy, which manages to obscure the costs of doing energy-related business from everyone *except the economy.* The true costs of SNG, LNG, and other uneconomic energy ventures appear in the form of inflation, unemployment, trade deficits, and environmental devastation.[9]

Hiding the costs of energy from our economy is an inferior substitute for a variety of policies, ranging from pricing energy at its full replacement costs to using overt subsidies. The advantage of even the latter over indirect protection has been eloquently spelled out by Peter Drucker.

> [A subsidy] can be limited in time—is indeed always limited in time, since public opinion and the legislature will eventually become impatient with any permanent subsidy. Politicians as well as bureaucrats dislike subsidies precisely because they are in the open. This, however, is their greatest virtue.[10]

For this reason, and others, EDF has opposed rolled-in pricing and declining block rates for gas and electricity, especially where the marginal or final block of consumption is priced at a level below even the misleadingly low average price of energy. It is important to understand that what EDF endorses, marginal cost-based utility rates, has its exact parallel in the advocacy of deregulating new gas prices. which EDF also supported.[11] Together, these reforms can finally connect, in the economic sense, one end of the gas system with the other. Without them, we must perform our autopsy on two different bodies, in two different regions, with two sets of very blunt instruments, and with predictably useless results. Much that is mistaken in analyses of the gas industry stems from a failure to look at its entirety.

Why is it that the United States shuns $2.60 per Mcf Mexican gas while eagerly doing business with its old friends, Algeria and Indonesia, to bring in their $4.50 per Mcf LNG? Perhaps the U.S. fears that Canada will let the roof off its $2.16 per Mcf gas. Surely the Canadians know what the United States pays for LNG and SNG. And if they do not, they may get an idea when Americans set up a $5–$7 per Mcf coal-based gasworks next door in North Dakota.[12] Or is it possible that Canadians (and, for that matter, the Mexicans) know what their gas is worth, while some American gas utilities have forgotten? What is it worth, anyway?

The Wisconsin Public Service Commission decided to find out. With a modest surplus of gas, Wisconsin's regulators held an auction among larger gas users in several utility service territories. They found out just what they expected and, of course, what folks in Texas already knew: gas is worth the value of the cheapest available alternative. When the spot price of gas exceeds this limit, the demand for it falls precipitously. Thus, as Wisconsin raised the price of its gas from roughly $2.50 to $2.80 per Mcf—the range of alternative fuel costs at the time—gas users switched to other sources. The first group switched to cheaper, heavy fuel oils. As the gas price rose, others started using more costly distillate oil. These are, of course, the very near-term alternatives to gas. In the long run, consumers could choose from many more alternatives, including conservation, coal, and new technologies that extract low- and medium-heat content gases from coal (which is much more economical than high Btu gasification).

Texas provides us with good empirical evidence concerning the value of gas, as the state's residual fuel oil and unregulated industrial gas prices have tracked each other closely. Large gas users there are leading the wave of conversion to coal and the investigation of new coal utilization technologies. Texas Utilities and its chief operating officer, T. Louis Austin, spearheaded this shift and, as Austin is quick to point out, Texas lignite deposits were acquired and developed on the basis of comparative prices, not regulatory mandates, when gas prices began to rise years ago.

The Wisconsin experiment and intrastate market experience tell us that there is an upper limit to the value of gas. This flies in the face of what some bureaucrats, gas utility managers, and investment bankers—eager to retain their utility customers—would have us believe when they propose to bring us $4.50 per Mcf LNG, $5.00 per Mcf SNG, or $6.00 per Mcf coal gas.[13] Spending more money for gas results in not being able to buy greater volumes of alternative fuel or gas itself at a lower price, whether they should be available now *or* in the future. In other words, investing in $6.00 per Mcf coal gas carries with it an opportunity cost—the cost of not being able to buy over twice as much $2.60 per Mcf Mexican gas, for example. Amazingly, no one asks whether disguising a several-fold increase in domestic gas prices is a sensible strategy for improving our balance of payments and energy independence positions, especially when our greatest concern is inflation.

The value of natural gas should be the foundation of any legislation designed to phase out regulation of wellhead gas prices. Indeed, an understanding of the limits of what producers could charge users for gas if they dealt directly with each other would considerably benefit the nation (as would a better appreciation of the economic consequences of overcharging). Gas is simply worth the cost of meeting unfulfilled demand for a sufficient number of present or prospective fuel users or conservers to bring local demand and supply into equilibrium. In the relatively near term this means that, at the wellhead, gas has a value equal to the cheapest imported fuels in markets where the gas is ultimately sold (minus any additional transportation costs).[14]

By disguising gas prices, we manage to pay a great deal more than we should for not having enough. For example, were gas pricing policy to reflect the actual opportunity costs of gas (the costs to the economy of not having enough), industrial and commercial users would switch fuels to some extent, and all gas users would seek to conserve. This would permit gas to become available to new customers on a price competitive basis. And, gratifyingly, it would end the exponentially growing practice of substituting attorneys for technology, a strategy strongly encouraged by current gas allocation schemes.

Instead, we penalize economically sensible gas conservation by imposing an allocation schedule that rewards those least prepared for emergencies and that seems geared to the concept of conservation through altruism. For this we pay by not having the proper fuel mix when cold weather creates shortages and by switching new residential consumers to electric space and water heating—an alternative that costs an equivalent of roughly $5.00 to $12.00 per Mcf for gas even after adjustments for fuel utilization efficiency are made. We also pay be encouraging utilities to find grossly uneconomical gas supplements, such as LNG, SNG, and high Btu gas from coal.

Suppose a gas utility increasingly invests in costly gas supplies that meet no market test. Hastened by increasing prices for conventional gas, the average price will soon exceed the cost of alternatives for a growing number of large gas users. They will switch fuels, perhaps abruptly, and the remaining gas customers will be saddled with extraordinarily high costs. As the consumption of gas swiftly shrinks, the fixed costs of its supply are spread over a much diminished volume of consumption. This may seem like extraterrestrial analysis in a time when all we hear about are gas shortages, but there is growing evidence that this process has already started in California, precisely where average gas costs are high and heavy oil alternatives are cheap.[15]

Prophylaxis

As we have seen, substantial markets for natural gas could disappear, perhaps within a decade. Such an outlandish outcome is strongly favored by prevailing gas rate designs, gas utility investment policies, and pending federal legislation, none of which is founded on the market value of gas. The pipedreams of the utilities, the gas pains of their customers, the claims of high administration officials to have traveled to the center of the earth and found no gas, and the advantage of those fortunate enough to possess sizable gas reserves discovered with yesterdays investments (and tax incentives) must be held in check if we are to avoid placing our entire economy on the coroner's table. At the same time, those who seek new gas must be given an incentive that reflects its value to fuel users—a price that would result were the market to clear. The problem of defining "new" gas is not insurmountable given present analytical techniques and proper positioning of the burden of proof. As we move into deep drilling, new frontiers, or areas of vast resource base but entirely unconventional geology, the problem further abates.

Too high a price would promote uneconomic investments (as the utilities eager to supply LNG, SNG, or coal gasification have so amply demonstrated). Too low a price will fail to reveal what we *must* know if our energy policy is to become rational: how much gas can we produce domestically at prevailing world energy prices. If insufficient gas is forthcoming, then (and only then) a commitment to higher cost alternatives will make sense. If an abundance of gas starts to flow, we will have regained a measure of autonomy, perhaps even to the point of influencing world energy prices. Regardless of how much gas can be produced, we will have obtained vitally needed information at the lowest cost.

Regulation at the state and federal levels must move to economically link production and consumption of gas. Key features of this shift include:

1. ending rolled-in pricing
2. eliminating declining block utility rates
3. shifting prices upward for the tail blocks of all consumers until these rate levels reflect marginal gas costs
4. instituting summer-winter price differentials for all consumers wherever indicated by either commodity or capacity costs and
5. substituting market incentives, such as use entitlements that can be traded, in place of rigid allocation schemes for as long as shortages persist.[16]

A good rule to remember is that user prices should reflect producer prices or opportunity costs, whichever are lower, if the economists' prescription for efficiency—marginal cost pricing—is to be adhered to. This means that the intermediaries must be price regulated *as if* they were common carriers; that is, as if users were buying gas directly from producers. To do otherwise will lead to paying too much money for too little gas and, assuming there is a good deal of gas yet to be found without resorting to utility schemes for finding it, to prematurely drying up the gas market for producers.

Finally, a note on taxation. If the energy industry is growing both more capital-intensive *and* less productive of capital (and there is evidence that this is the case), we will have to rethink and reformulate tax policies that were geared to an earlier era, when capital productivity was increasing. As I hinted earlier, one structural reform would be to peg incentives for capital formation to the real productivity of capital investment. Another would insure that the tax laws are not undermining economically attractive alternatives to the conventional plans of utilities or producers. For example, tax credits for solar energy installations—currently in a state of technological infancy[17]—should, at the very least, reflect the tax consequences of *not* resorting to conventional fuels. Even better, of course, would be to eliminate *all* tax incentives and reduce tax rates.

Conclusion

There are certain fundamental principles that must be the foundation of a successful energy policy. Sadly, the attention of the nation's policy-makers is focused elsewhere. For example, the debate over deregulating natural gas producers' prices has focused too much attention on a single thread of the energy policy fabric and has ignored the fact that every pipeline has two ends. It is silly to concentrate on the economics and technology of one end—gas producers—and not take a hard look at the other end—gas consumers. Indeed, the producers are in for a rude shock if utility pricing policies abruptly collapse the gas market.

Rolled-in pricing and open-ended tax subsidies that so perversely obscure the cost of energy must end. After all, a consumer can save money by using less energy; he cannot reduce his tax bill—part of which subsidizes energy production—in the same way. Regardless of whether our present energy circumstances are called a crisis, a shortage, or a hoax, new policies regarding energy are far more important than most consumers recognize.

The energy situation that has attracted so much attention will soon appear in other sectors of our economy, most notably, raw materials and water. Thus, the policies accepted for energy will probably provide the keystone for reforms in other areas. In a very real sense, a mistaken energy policy is a prescription for national disaster, not only for energy, but also for other natural resources.

Notes

1. Nothing strikes today's adherents of conspiracy theories more strange than stories of pipeline companies charging producers to take away their unwanted gas pursuant to the Texas Railroad Commission's early "no flare" orders. As we shall see, the interests of the pipelines and the producers are still dissimilar, more so than many suspect.

2. I am indebted to a friend and former colleague, Philip J. Mause, now practicing law in the District of Columbia, for pointing out several of these analogies. In a recent letter to a gas industry acquaintance, Mause noted that: "The first issue was whether taxes should be imposed upon whales beached on the shore through natural causes. These whales provided a windfall to the finder. On the other hand, a tax imposed upon all whales would reduce the incentive to hunt whales off-shore. Thus, an early English Governor of New York hit upon the device of a selective tax to be applied only to whales washed up on shore. It was quickly discovered that administrative problems arose because whale finders alleged that they had actually harpooned the whales at sea and brought them ashore. The result was an administrative nightmare,

serious administrative costs, and a certain amount of corruption. In addition, many Long Island whalers shipped their oil to New England to avoid the tax."

3. Unless otherwise noted, these historical vignettes are drawn from Elmo Paul Hohman, *The American Whalesman* (New York: Longmans Green, 1928), pp. 25–36. See also Alexander Starbuck, "History of the American Whale Fisher from Its Earliest Inception to the Year 1876," *Report of the United States Commission of Fish and Fisheries for 1875–76*, vol. 4 (Washington, D.C.: U.S. Commission of Fish and Fishers, 1877). The twentieth century demise of whaling, perhaps one of the most striking examples of inept management of a replenishable resource, is well-reported by George L. Small in *The Blue Whale* (New York: Columbia University Press, 1971).

4. Frederick P. Schmitt, *Mark Well the Whale* (Port Washington, N.Y.: Kennikot Press, 1971), pp. 8–9. In oil industry language, whose color has faded little over the past 140 years, Hunt went on to insist that the *Monmouth*'s owners "abandon her to the Government, let her masts rot through her bottom, and moor her where she lays, until the Day of Judgment—of the United States Court; and if there is a particle of independent honesty left in that court, force the government to disgorge the full amount of damages, created by a gang of *political black-legs.* . . ." But the government prevailed and the duty was paid.

5. Here, of course, our entire energy policy comes unglued. Many environmental groups have long argued that energy be priced at (*not* above) marginal costs precisely because they have faith in the innovative capacities of our economy. Gradually, EDF has won increasing numbers of electric utility managers and regulators to its side, but these are still a minority. Few if any of these cases have attracted the attention of the media, so a sizable portion of the EDF's actions is virtually unknown.

6. Peter Drucker, *The Age of Discontinuity* (London: Pan Books, 1971), pp. 85–86.

7. Perhaps the foremost cause of the present federal legislative inaction (and of past policy failures as well) has been the repeated attempt to solve *both* equity and efficiency issues, which stem from rapid energy price increases and from inflation, with a *single* policy instrument unencumbered with any appreciation for the importance of regulation at the state level where prices for half the energy we use are established. Indeed, in their capacity to oversee our gas and electric utilities, state regulatory authorities establish prices and review investment policies for nearly 70 percent of our nontransportation energy supplies.

8. The Petrochemical Energy Group (PEG), representing many naphtha and propane consumers in industry and agriculture. Here, the interests of large and small users are indistinguishable, however, since rolled-in pricing extracts subsidies from all existing customers to pay for gas supplies for new users or those whose use is avoidable at lower cost.

9. One strategy superior to producing SNG from naphtha would be end-user fuel switching from gas to naphtha. This has the dual advantage of

less capital investment and higher thermal conversion efficiency—as long as natural gas is used for industrial heating applications or new attachments. See E. R. Habicht, Jr., testimony before the Federal Energy Administration on proposed synthetic natural gas feedstock regulations, Washington, D.C., September 2, 1977.

10. Drucker, *Age of Discontinuity,* p. 89. Drucker points out that "protection creates dependence which is increasingly difficult to abolish."

11. U.S., Congress, House, *Committee on Interstate and Foreign Commerce, Subcommittee on Energy and Power,* 94th Cong., 2d sess., May 18, 1977 (testimony of E. R. Habicht, Jr.).

12. Perhaps the most ridiculous public or private statement on energy since the fall of 1973 was made by Secretary Schlesinger. He said that, indeed, the Canadians would take notice of higher Mexican gas prices and act accordingly. So what? Would not the Canadians take equal notice of unfulfilled gas demands in the absence of Mexican purchases and likewise act accordingly? Such gratuitous insults to a friendly neighbor's acumen are unseemly.

13. People who espouse such uneconomic goods are in one way like the "Coneheads," a family of mother, father, and daughter who appear in skits on a late-night televised comedy show. The Coneheads, possessed of decidedly peculiar habits, have been sent to earth from another civilization on a mission that depends on their becoming assimilated into society. The trouble is, *they have forgotten their mission.*

14. Note that there are very real differences in both the cost of alternative fuels in different markets and in the spare capacity to deliver gas to different regions. This explains why the Northeast gas market may command higher prices than the far more heavily industrial Texas market, as well as why Canadian gas may be worth somewhat less than Mexican gas, at least during the winter heating season.

15. See, for example, I. Y. Borg, C. J. Anderson, R. Sextro, and B. Rubin, "An Overview of Recent Trends in California Natural Gas Consumption (1975–1977)," UCRL–52498. Lawrence Livermore Laboratory, June 15, 1978.

16. Wisconsin, Public Service Commission, docket no. 6680–GR–3, October 1977 (testimony of E. R. Habicht, Jr.). Under no circumstances should tradeable use entitlements be confused with current federal entitlement programs. The idea of tradeable use entitlements, first set forth by California Institute of Technology economist Roger Noll, is based on the concept that market-determined gas use priorities are superior to priorities established by an inflexible regulatory authority.

17. See E. R. Habicht, Jr., "Electric Utilities and Solar Energy: Competition, Subsidies, Ownership and Prices," in *The Solar Market: Proceedings of the Symposium on Competition in the Solar Energy Industry.* (Washington, D.C.: Bureau of Competition, Federal Trade Commission, 1978).

The Policy-Induced Demand
for Coal Gasification

by Richard L. Stroup

This paper begins with two hypotheses. The first is that coal gasification in the United States will not be economical for commercial purposes for the remainder of this century. The second is that regulatory distortions will probably cause us to buy commercial quantities of synthetic gas (SNG) from coal within 10 to 15 years.[1]

This statement was written in 1975. There is evidence today supporting both hypotheses. To explain the persistence of this inefficient energy supply strategy, this paper traces the various perverse impacts of governmental pricing policies whose declared intent was to prevent economic inefficiencies and to benefit gas consumers.

For some years, natural gas price regulation has been at the center of energy policy debates. Indeed, one can argue that regulation induced gas shortages have intensified every symptom of the so-called energy crisis. Electricity blackouts are promoted by energy demands that would otherwise be served by natural gas produced (or not used by current consumers) at higher gas prices. Regulation also inflates oil import demands and reduces the supply from domestic oil wells. Gas from marginal oil wells is currently burned off because the suppliers cannot profitably market it at the controlled price.

A related problem is the trade-off of environmental quality that will result from prematurely exploiting the vast energy potential of

The author wishes to thank the Scaife Family Charitable Trust and the Montana Agricultural Experiment Station for support during the preparation of this article. Appreciation is also due to Lloyd Orr and Henry Goldstein for their helpful comments. The author retains responsibility for the contents of this article.

American coal and oil shale. Sophisticated environmental groups are among those who recognize the strong connection between the gas pricing issue and environmental trade-offs. The Environmental Defense Fund (EDF), for example, has intervened in natural gas rate cases to encourage the development of marginal cost pricing. This plan would require each user to pay the full system cost of his own gas consumption and to receive the full benefit of conservation.

It may surprise some people to see environmentalists and market-oriented economists using nearly identical arguments based on internalizing costs and benefits to decision makers, but any fight for a socially efficient policy is a positive sum game. For example, even one who is ideologically opposed to windfall profits may accept some increase in price if consumers as a group will gain. Ideological differences are often overcome under such circumstances. Institutional arrangements that hinder mutually beneficial exchanges also increase inefficiency. These institutions can be remolded, however, and as the gain from doing so increases, we have some hope—indeed some evidence, such as EDF actions favoring gas deregulation—that market economists and groups generally seeking more governmental regulation can in fact reach agreement on some important issues.

Evidence that coal gasification is inefficient ranges from data on the price responsiveness of demand for natural gas to estimates of the supply responsiveness for other energy sources. Whether measured by energy inefficiency, water and environmental costs, or simply in dollar values, the high cost of SNG strongly indicates that a socially efficient energy strategy would not employ its commercial production. This will likely be true for several decades. Nonetheless, movement toward SNG production with existing technology is proceeding as part of the federal energy strategy.

SNG: The Product and the Production Process

Coal gasification technology dates from 1670,[2] but the most thoroughly proven technology is probably that used in Lurgi units, first developed in Germany over fifty years ago. The Lurgi process combines crushed coal with steam and oxygen under high pressure to make a mixture of hydrogen, hydrocarbons, nitrogen, and carbon oxides. Before this mixture qualifies as pipeline quality gas, it must undergo methanation, a commercially unproven process that removes all carbon monoxide and some carbon dioxide and raises the heating value from less than 500 to 900–1,000 British thermal units per standard cubic foot (Btu/scf). Although newer technologies are being explored, the first plants are expected to use the Lurgi process with methanation.[3]

SNG is a highly desirable product, since it is easily handled and burns cleanly. It is environmentally less objectionable than most other fuels, including the coal from which it is made. SNG production, however, is an entirely different matter. Despite a growing knowledge about environmental and health dangers from the commercially untried technologies that could produce SNG, much ignorance remains. It does seem clear that a number of harmful chemicals, including many suspected carcinogens, would be present in the effluent and product streams and the fugitive emissions from any future SNG plant.[4]

The Rationale for SNG

Despite high private and social costs, plans to produce SNG persist. The most direct reason for this is the current shortage of natural gas. SNG is a direct substitute for natural gas, though it often has slightly less energy content per unit volume. The amount of gas demanded exceeds the amount supplied, and the shortage has grown over the last few years.

This shortage began in 1971. Its cause, however, dates from the 1950s, when the Federal Power Commission began regulating well-head natural gas prices.[5] While natural gas has a high value, especially to home consumers, due to its clean and continuously deliverable nature, the Federal Power Commission (now the Federal Energy Regulatory Commission) controlled its price below a market-clearing level (where the amount produced equals the amount demanded). Until 1970, new wellhead prices were below $.20 per thousand cubic feet (Mcf). By comparison, electricity generated from coal, oil, or natural gas plants costs ten to thirty times as much. Even today, the wellhead price of interstate gas is less than one-third the cost of electricity at new plants. Prices controlled at this level understate the value of gas and, consequently, increase the amount of gas that consumers demand and retard incentives to produce and explore for new sources of natural gas. The *structure* of regulated rates also contributes to greater quantities of gas being demanded. A "declining block" rate structure prevails, so that the average cost of *extra* gas used by a customer declines as the customer uses more gas.[6]

Since gas is clean and convenient, a shortage naturally suggests supplementary supplies. At this point, regulation of gas prices charged by pipeline distributors becomes important. Although initial

SNG supplements will probably cost close to $5.00 per Mcf, this cost is likely to be "rolled in" with cheaper natural gas. So users of the new gas will pay only about half its cost, while the other half will be paid from the higher price on all natural gas in the system.

For example, the current federally regulated ceiling on domestic natural gas sold in interstate markets is below $1.50 per Mcf at the wellhead. Consider a gas distributor who bought 80 percent of its gas at $1.40 and 20 percent SNG at $5.50 per Mcf: the average gas cost would be $2.22. If the declining block rate structure of this distributor placed operating costs (other than gas costs) on the first few Mcf's per month purchased by each customer, most people would face a price of $2.22 per Mcf on additional gas purchased (or saved). The importance of rolled-in pricing to the success of SNG sales becomes apparent. The buyer can pay $2.20 for gas that costs the distributor (and, therefore, its customers as a group) $5.50. The customer who uses an extra Mcf of gas pays less than half its cost. The rest is spread among all other customers of the utility.

The combination of federal wellhead regulation and state regulation of distributors means that expensive SNG can be used as though it were cheaply produced. Its high cost is not the deterrent to the production of SNG, which efficiency would demand. Only in this sort of price-distorted environment would a firm seeking profits want to invest a billion dollars to produce a product the revenue from which is not expected to cover its cost of production. Yet, this subsidy of SNG by natural gas users appears likely.[7]

Probably less recognized are the distortions and asymmetric effects that rate of return regulations can be expected to have on the regulated firm's view of production costs versus delivery uncertainties. A firm with a regulated maximum rate of return will presumably avoid risky ventures, even if the expected rate of return is well above both normal and regulated rates. Large losses are likely to be borne at least in part by the firm; yet large gains would clearly not be available to them should the project be a big winner.

Small gambles may be acceptable, however, at least on the production cost side. A firm considering an SNG project, for example, may not be dissuaded by the risk of, say, a 50 percent overrun on plant costs, since the regulated rolled-in price of the product could probably be raised by enough to cover the extra cost, avoiding any loss to the utility (assuming that the firm's gas demand is relatively insensitive to a higher price, in the regulated context). The same degree of risk to society on the output side might be unacceptable to the firm. A utility that cannot deliver part of its planned output will probably not be paid for that output. Making other customers pay extra to cover fixed costs plus an allowed return on unused distribution capacity built to serve customers cut off in a shortage may not be acceptable to the regulators, so this type of loss may be borne by the utility itself. Even if every customer is largely protected from the risk of shortage by the ability to switch temporarily (or permanently) to, say, coal or fuel oil, this socially minor risk is seen as large by the gas utility.[8]

This distorted view of risk may be crucial to the question of SNG efficiency as seen by the utility. If natural gas were available in sufficient quantities at $2 per Mcf while more certain delivery of SNG costs $5.75, the utility might still want to choose SNG. *In their view,* the higher costs would likely be offset by the ability to pass costs to the user. Yet the social benefit of smaller cost, associated in this hypothetical case with natural gas, would not be appropriable by the distributing firm, whose rate of return cannot exceed a specified figure.[9] Of course, if everyone involved, including wise, unbiased regulators, knew these facts, only the natural gas option would be allowed. The point is that the firm does have distorted incentives, and the regulatory commission may well be less than perfectly informed or perfectly unbiased.

In brief, the demand for SNG is caused by the shortage of cheaper natural gas brought on by wellhead price regulation. The shortage is aggravated by declining block rate structures. Both the possibility of rolled-in prices (a form of cross subsidy) and the distorted view of risk (caused by rate of return limits) increase the gas distributors' desire to produce SNG. These impacts are, of course, in addition to the subsidies already given to SNG process development efforts and proposed loan guarantees for commercial plants.

Cost Comparison

In any cost consideration of synthetic coal gas, it should be kept in mind that pipeline-quality synthetic gas production is not a commercially proven technology. Further research and development need to be funded and carried out before the first plant operates. For example, significant amounts of development are required in high-pressure feed systems,[10] pollutant analysis and control,[11] application of technology to various coal types,[12] the methanation process,[13] and scaling-up processes from experimental to commercial size.[14]

In addition to these more scientific areas, work is needed on engineering problems of actual plant construction. These include un-solved problems of field erection and fabrication of pressure vessels.[15] The large size of such vessels precludes shop fabrication and necessi-tates field adaptions of previously untried precision techniques.

Most independent estimates of coal gasification costs have re-cently run higher than $4 per million Btu's. Perhaps the most optimis-tic estimate has been the $3.40 per million Btu's estimated by the backers of Synthane, a new experimental process.[16] A more recent and probably more reliable figure for SNG costs would be the $5.83 per million Btu estimate, net of transmission and distribution cost, for a Lurgi methanation unit proposed in July 1978, for North Dakota.[17] The Lurgi process is well tested, though the required methanation step is not.

By contrast, the MIT energy simulation model estimates that a new contract field price of about $2.26 per million Btu's would clear the natural gas market.[18] If this estimate is even close to being correct, deregulation would end any competitive chance for SNG in the near future. The proposed North Dakota SNG plant would, by its propo-nents' estimates, produce gas costing more than two and one-half times the estimated market-clearing natural gas price. Another MIT report indicates that electricity from coal (or other sources) would also be cheaper than SNG when used with modern heat pump technol-ogy.[19] Of course, coal burned directly for heat would be much cheaper than SNG, reflecting in part the increased thermodynamic efficiency of avoiding the chemical transformations required in gasifi-cation. The SNG processes, inefficient in terms of economics, ther-modynamics, and the environment, simply could not stand careful scrutiny were it not for direct and indirect governmental interference.

Conclusion

With the declared intent of protecting natural gas customers from high prices to monopoly pipelines and gas distributors, state and fed-eral agencies established regulatory control over those firms' prices. Price controls were extended to interstate gas at the wellhead by the Supreme Court. Gas prices held below equilibrium had the predict-able effect of increasing quantities demanded while reducing quan-tities discovered. Shortages of gas followed. Intensified by declining block rate structures, these shortages threaten to burden consumers with extremely costly supplemental gas, such as the environmentally hazardous production of SNG from coal. The likelihood of this burden's onset is increased by the use of rolled-in pricing and the distorted view, which regulation causes utilities to have, of the cost-risk trade-off. Subsidies to SNG from the federal purse strengthen the threat. The existence of several serious environmental risks is well established, though the risks are not well quantified at this time.

It is ironic that collective actions taken presumably to correct failures in a purely volunteer or market setting are now seen to cause serious inefficiencies (resources with high replacement costs used as if they were cheap), inequities (a user can use more and shift much of the cost to others), and increased environmental burdens (more oil and coal are used in place of cleaner, but wasted and undiscovered gas). The most direct way to attack these problems would be to eliminate the collective actions causing them and to rely on voluntary exchange. Another approach would be to change public policies to try to mimic the market, keeping regulation in place.

In addition to eliminating the demand for SNG, a return to the free market in natural gas at the wellhead would result in higher natural gas prices, no gas shortages, and a reduced demand for electricity, oil, and coal. Owners of residential gas hookups would, of course, be made worse off by the higher burner tip gas price.[20] Note, however, that many *poor* people do not own their own homes. For them, at least part of the value of low gas prices in their homes or apartments is captured by the landlords via higher rents as compared to housing units with electric or oil heat. Clearly, energy users forced to use the latter, more expensive energy forms due to gas shortages (or for other reasons) would be better off without wellhead gas price regulation since demand, new production, and prices of these alternatives would fall. Breyer and MacAvoy argue persuasively that deregulation on balance would help residential energy consumers *despite* higher profits to gas well owners.[21] Public policy makers should probably be reminded that "profit" is not a four letter word and free exchange is normally a positive sum game.

To eliminate regulation of pipeline transmission and distribution prices would produce results more difficult to analyze. Some elements of natural monopoly exist here, and the net impact on consumers is not so obvious, though cross subsidy of SNG and the problems of marginal price well below marginal cost would surely disappear.

Federal subsidies to SNG commercialization would have to be stopped using political means. Since there are no apparent positive technological externalities to SNG production, the argument for subsidies is strictly redistributive. As indicated above, higher wellhead natural gas prices are a far cheaper way to get more gas to consumers. Yet the federal treasury, largely a common pool resource, is always a tempting target for those whose jobs, business profits, or regional development can be aided by federal money. The outlook is not cheerful.

To try to mimic the market by eliminating rolled-in pricing and declining block rates would undoubtedly improve regulation. Note, however, that even these improvements would leave some room for potential gains. The *total* utility bill is still generally not a reflection of total opportunity cost. The revenue constraint logically restricts the efficiency of the system. Development of a transferable "entitlement" for the customer to buy known gas reserves at an artifically low price could partially solve this problem. The confiscation of well owners' wealth continues as now, but the buyer would always face the true opportunity cost of his consumption if he fails to sell his entitlement. Even here, the supply side of the problem is distorted by producers' (and searchers') failure to receive the production inducement of full market value.

Any attenuation of private property rights is likely to be costly, whether by price regulation or other means. Either the incentive to produce for the benefit of others is reduced, or the incentive to conserve resources is reduced. Often, both occur. Further, since individual responsibility is at least partly curtailed, individual authority to act (freedom) must also be limited. Unconstrained private rights are, of course, demonstrably imperfect in predictable circumstances. Monopoly, externality, and public goods problems all argue for the consideration of collective action. The result of collective actions on SNG production probabilities, however, strongly indicates that "corrective" collective action can create far worse problems than the problem it attempts to solve. Although protected from paying higher prices to gas well owners and higher profits to distributors, gas users face both shortages *and* the likelihood of being required to pay for high cost SNG.

Notes

1. R. L. Stroup and V. House, "The Political Economy of Coal Gasification: Some Determinants of Demand for Western Coal," staff paper 75–17 (Bozeman: Department of Agricultural Economics and Economics, Montana State University, 1975).

2. For a history and description of coal gasification technology, see U.S. Department of Interior, Office of Coal Research, *Evaluation of Coal Gasifica-*

tion Technology, Part I: Pipeline Quality Gas (Washington, D.C.: Government Printing Office, 1973).

3. This paragraph and certain other passages are taken or adapted from R. L. Stroup and W. N. Thurman, "Will Coal Gasification Come to the Northern Great Plains?" *Montana Business Quarterly* 14 (Winter 1976):33–39.

4. A discussion of these problems is found in Stanley M. Greenfield, "Environmental Problems with Fossil Fuels," in *Options for U.S. Energy Policy* (San Francisco: Institute for Contemporary Studies, 1977), pp. 92–103.

5. A history of the Federal Power Commission is found in S. G. Breyer and P. W. MacAvoy, *Energy Regulation by the Federal Power Commission* (Washington, D.C.: The Brookings Institution, 1974).

6. An excellent discussion of this problem is found in P. J. Mause and J. H. Bailey, "Brief of the Environmental Defense Fund, Before the Public Service Commission of Wisconsin," mimeographed, 6680–GR–3 (Madison, Wis., February 6, 1978).

7. R. E. Boulanger, "Additional Prepared Testimony, Before the Federal Energy Regulatory Commission," Docket nos. CP78–391, CP75–278, CP77–556 (Washington, D.C., July 14, 1978).

8. It is important to note that wellhead price regulation has not only reduced the quantity of gas supplied by lowering the real price of gas, but also by increasing the uncertainty in its selling price even *after* a contract for delivery is signed at a specific price. This has happened because federal regulators have retroactively revised private contracts, requiring millions of dollars in refunds from producers to distributors. Sanctity of contract is unattainable here, reducing the supply of gas.

9. It also seems reasonable to assume that the purchase of even a very expensive insurance policy, paid by users, against the risk of nonavailability of gas, does not go against the desires of decision makers in the utility bureaucracy, whose management problems seem very likely to be reduced thereby.

10. U.S. Department of Interior, Office of Coal Research, *Evaluation of Coal Gasification Technology*, pp. 23, 39, 44–45; James R. Garvey et al., *Final Report of the Supply-Technical Advisory Task Force: Synthetic Gas-Coal* (Washington, D.C.: Federal Power Commission, 1973), p. VI–2.

11. U.S. Department of Interior, "Synthetic Fuels from Coal," Task Force Report of the *Project Independence Report* (Washington, D.C.: Government Printing Office, 1974), pp. 23, 83–90; Garvey et al., *Final Report of the Supply-Technical Advisory Task Force.*

12. U.S. Department of Interior, "Synthetic Fuels from Coal," p. 24.

13. U.S. Department of Interior, Office of Coal Research, *Evaluation of Coal Gasification Technology*, pp. 26, 44; Garvey et al., *Final Report of the Supply-Technical Advisory Task Force*, pp. VI–4, VI–5.

14. Department of Interior, "Synthetic Fuels from Coal," p. 24.

15. Ibid., pp. 63, 68–70, 118.

16. This figure was reported in the *Energy Users Report*, no. 27 (December 15, 1977):29. Note that backers of the process have every incentive to

report optimistic cost figures to increase the likelihood of future funding. Also, any pessimists on such a team have every incentive to find other, more secure work.

17. Boulanger, "Additional Prepared Testimony," exhibit REB–28.

18. This figure is adapted from Pindyck, using his implied inflation rate of 6.5 percent, to estimate the cost in 1978 dollars. See Robert S. Pindyck, "Prices and Shortages: Policy Options for the Natural Gas Industry," in *Options for U.S. Energy Policy.*

19. See O. Hammond and M. Zimmerman, "The Economics of Coal-Based Synthetic Gas," *Technology Review,* July-August 1975.

20. This assumes that SNG and other supplemental supplies do not push the average cost of delivered gas higher than the deregulated gas price. The assumption may or may not hold. If not, regulation looks even worse.

21. Breyer and MacAvoy, *Energy Regulation.*

The Navajo and Too Many Sheep:
Overgrazing on the Reservation

by Gary D. Libecap and Ronald N. Johnson

It is perhaps ironic that the Navajo, a people so closely identified with nature, should engage in persistent and severe overgrazing of their lands. Indian Service records indicate that stocking beyond the capacity of the range began on some parts of the reservation as early as 1890. By the 1920s the effects of overgrazing, wind and water erosion, and depletion of the quality of the land had spread throughout the reservation. The Indian Service implemented a forced stock reduction program in the 1930s and 1940s that succeeded in reducing herd size to the carrying capacity of the range, which was estimated at 512,000 sheep units. Yet by the 1960s total stocking had rebounded, accompanied by a return of overgrazing. In this paper, overgrazing is defined as that level of use that would persistently deplete the productive capacity of the range. Following the procedure used by the Navajo tribe and the Department of the Interior, we measure it by the ratio of animal stocking to carrying capacity.[1]

For the Navajo, raising livestock is the traditional means of earning a living. Although there has been a decline over the years relative to other sources of income, the pastoral economy remains an important source of earnings for many Navajos. Historians, anthropologists, and sociologists have extensively studied subsistence stockraising and the vital role of the pastoral economy in the cultural and religious beliefs of the Navajo.[2] They have neglected, however, to emphasize the importance of property rights arrangements in explaining land use. In this paper, we depart from the emphasis placed

An earlier version of this article appeared as "Legislating Commons: The Navajo Tribal Council and the Navajo Range," *Economic Inquiry* 18, no. 1 (January 1980):69–86.

on cultural and religious values and explain the occurrence of over-grazing using the economic theory of property rights. This is not to deny those values, but to assert that institutional constraints are more likely to explain resource use. The role of property rights institutions in explaining individual behavior has been outlined by Demsetz, Alchian, Cheung, and others.[3] They argue that income and production can be maximized over time only if control over resources is clearly established, enforced, and transferable. Such control depends on the coercive power of some agreed upon authority.

We argue that overgrazing is the result of a failure by the Department of the Interior and the Tribal Council to enforce property rights and to adopt an enforceable grazing program. We further show that the problem of overgrazing is exacerbated by the policy of breaking up large herds into small ones, thereby raising the costs of negotiating, transferring, and enforcing property rights. We detail the nature of range land use and the role of the Indian Service (later the Bureau of Indian Affairs) and the Tribal Council from 1930 to 1978. In addition, we provide testable implications regarding the relationship between herd size, overgrazing, and fencing.

Property Rights and the Roles of the BIA and the Tribal Council

Overgrazing on the Navajo reservation has been a continuing problem since early in this century. To understand how property rights arrangements and the actions of the Department of the Interior and the Tribal Council contribute to overgrazing, institutional changes must be described. This section, then, is broken into three parts: (*a*) a description of property concepts that have remained more or less intact; (*b*) an outline of the increased involvements of the Indian Service during the stock reduction of the 1930s and 1940s; and (*c*) an account of the shift to tribal control in the 1950s.

General Property Concepts

Navajo tribal members individually own the animals they graze, and such rights are transferable. They do not, however, hold title to the surface land, but instead hold usufruct or use rights.[4] Actual title is held by the federal government, and administration of the land has shifted from the Department of the Interior to the Tribal Council. The land area grazed by any family is referred to as a customary use area, a name that reflects the Navajo practice of allocating land on the basis of prior appropriation and use. Since maintenance of use rights requires grazing, the amount of land in any customary use area depends on the number of animals grazed and the land quality. Failure to fully utilize the range within this area invites trespass. Sheep and goats are grazed within the customary use area, though cattle and horses are allowed to roam beyond the area's boundaries. Actually, those boundaries have never been precisely defined in the law; neighboring herders establish them through informal agreements. Until the 1930s, there was little involvement by the Department of the Interior or the tribe in grazing matters. Unlike their neighbors, the Pueblos, the Navajo had no permanent Tribal Council until the early 1920s, when the Commissioner of Indian Affairs set up a Navajo Tribal Council to confirm oil leases in the northern or San Juan portion of the reservation.[5]

Range conditions, however, brought dramatic changes in federal grazing policy. The rapid growth of the Navajo population from 26,624 in 1910 to 40,858 in 1930 on a static land base intensified the grazing problem, and a general drought and the Depression compounded it. The federal government began various conservation and reclamation projects in the early 1930s, including the construction of Hoover Dam and the California Central Valley Project, the passage of the Taylor Grazing Act, and the establishment of the Civilian Conservation Corps.[6] Studies of the deterioration of the Navajo range and Secretary Ickes's fear that increased silting of the Colorado River would threaten Hoover Dam led to the adoption of a stock reduction program on the reservation.[7]

Property Rights and the Stock Reduction Program

The stock reduction program signaled the active involvement of the Department of the Interior in Navajo grazing activities. The Commissioner of Indian Affairs, John Collier, initiated the program in 1933 with the purchase of 86,517 sheep at roughly $2 per head, a price that closely approximated the $2.30 market price of Arizona sheep.[8] In that year, there were roughly one million sheep units[9] on land with an estimated carrying capacity of half that.[10]

From the beginning, the stock reduction program was the focus of an intense political rivalry between J. C. Morgan and the Northern (San Juan) Navajo and Chee Dodge and the Southern (Fort Defiance) Navajo. That rivalry no doubt affected not only the perception that individual Navajos had of the stock reduction goals but the effectiveness of the program as well, since the tribe's leaders were neither consistent nor uniform in their support of it.[11] Since the 1930s, grazing regulations have remained a political issue on the reservation, and tribal leaders generally support less enforcement rather than more.

The equity issue was a central element of the stock reduction program, and it affected not only how the reduction was planned but how property rights were subsequently assigned. For instance, in 1933 Collier argued that "every effort should be made to purchase stock from Indians owning large herds rather than small ones."[12] The Tribal Council supported this position by requesting that the reduction focus on herders with larger flocks.[13] In the early years, however, the program may have affected small herders more than large. Yet, by 1941 the issuing of sheep permits sharply reduced allowable herd size and the large flocks were broken up.[14]

It is important to note that the emphasis on equity has generally been applauded by those who have studied Navajo grazing conditions.[15] Unfortunately, this preoccupation with equity has meant that the efficiency effects of a policy aimed at small herds have been ignored.[16]

Collier ordered a second reduction in herd size in 1934, and 148,000 goats and 50,000 sheep were purchased at $1 and $2 a head, respectively.[17] The Tribal Council agreed to the purchases, but exempted those who held 100 or fewer sheep units.[18] Resistance to the sale increased, perhaps because Arizona sheep prices had risen to $3.35.[19] In 1935, Collier split the reservation into eighteen range management districts, determined the carrying capacity of each, and tried to reduce the animals in each area to that number. This led to a third reduction in 1937, when 70,939 sheep units were purchased (over half were cattle and horses). The last major purchase was in 1939 when the government acquired 10,000 horses.[20] Smaller acquisitions continued through 1946.[21]

In 1941, the Indian Service began issuing sheep permits that *formally* assigned grazing rights by specifying the number of sheep units each permit holder could run. Permits were assigned to family heads by district, while the carrying capacity determined the total number of units to be divided among the population. As before, equity was emphasized by reducing the share to large herders and freezing the number grazed by small herders.[22] The permits specified grazing districts, but neither property boundaries nor the actual area to be grazed was spelled out (fig. 3). Boundaries were to correspond with the customary use area held by the assigned permittee. They were neither surveyed nor marked and, accordingly, were deter-

NAVAJO TRIBE GRAZING FORM NO. 1-3-66-5M

(CENSUS NO.)	(NAME)	

Assigned Brand _____   District No. _____

Date Issued _____   Permit No. _____

                                                    Sheep Units

Horses Permitted _____ ,   Totaling _____   _____

Sheep and Other Livestock Permitted _____   _____

Total Permitted _____   _____

Season   This District _____
of Use
           Elsewhere and Dates _____

United States
Department of the Interior
Office of Indian Affairs

GRAZING PERMIT

Navajo Reservation
Window Rock, Arizona

━━ PERMIT CONDITIONS ━━

BY AUTHORITY of law and pursuant to the regulations in Part 72-Navajo Grazing Regulations, Title 25 C. F. R.-and amendments thereto, the above-named Indian is hereby granted
permission to hold and graze the number and kind of livestock as specified above on the Navajo
Reservation for the time and in the district or districts as stated above and thereafter until further notice, subject to compliance with the Range Management Plan for the district or districts
and any changes made in accordance with and pursuant to the said Grazing Regulations as
amended.

This permit shall not be assigned, sublet, or transferred except as provided in said Grazing
Regulations.

The Superintendent shall make desisions relative to the interpretation of the terms of this
permit and enforcement of Grazing Regulations.

Done at the Navajo Agency on this _____ day of

_____ , 19 _____

_____
Sub-agency Superintendent
[20 F. R. 2895]

Fig. 3. Grazing permit

TABLE 2. Maximum Permit Size by District, 1941

District	Maximum (*in sheep units*)	District	Maximum (*in sheep units*)
1	225	10	153
2	161	11	105
3	280	12	104
4	72	13	200
5	280	14	61
7	237	17	275
8	154	18	238
9	83		

Source: David F. Aberle, *The Peyote Religion Among the Navajo* (Chicago: Aldine, 1966), p. 67.

Note: Districts 6, 15, and 16 are not included because 6 was the Hopi Reservation and 15 and 16 were largely off the reservation proper.

TABLE 3. Percent of Permit Holders with Herds of 100 or Fewer Sheep Units

District	1943	1957	1978
1	52%	73%	86%
2	65	75	85
3	59	67	81
4	70	87	98
5	53	69	78
7	65	73	84[b]
8	66	71	88
9	100	91[a]	97
10	79	85	92
11	84	89	96
12	79	83	87
13	75	74	81
14	100	98	84[b]
17	66	73	84[b]
18	80	82	84[b]

Source: Authorized number of sheep units per permit from BIA records at Window Rock, Shiprock, Tuba City, Fort Defiance, and Chinlee.

a. Reflects issuance of special permits that allowed slightly larger herds
b. Fort Defiance Agency as a whole, individual district data not available

mined by informal agreements made among neighboring permit holders. This system has remained the basis for grazing rights on the reservation.

The Indian Service, reflecting the equity orientation, limited the number of sheep each permittee could hold. Table 2 shows the planned maximum herd size for each district in 1941. Those limits were relaxed later that year with the adoption of a 350 sheep unit maximum for each permit holder. Nevertheless, the result has been to make large herds into small ones (see table 3). Individual herd size has declined over time, in contrast to the pattern of non-Indian sheep holding in the Southwest, where herd sizes have grown.[23] In the *Navajo Yearbook,* Young estimates that at least 300 sheep are necessary for a minimum income of $3,000.[24] The policy of assigning permits to small herders, therefore, means that few can make a living from grazing unless there is costly negotiation and herd consolidation. Yet, this has not happened. Figure 4 shows the very general nature of the Bureau of Indian Affairs (BIA) transfer agreement, which specified only animal units and not land. The transfer, then, merely assigns sheep units within the district of the seller.

Discussions with BIA officials indicate that most exchanges (perhaps as high as 90 percent) have been through inheritance. Unfortunately, as the original permit is split up among heirs, the customary use area's boundaries become less clear. Once grazing shifts from the original permittees who had agreed to the borders to others who had not, a new definition of boundaries becomes necessary. Interviews with tribal and BIA range personnel and with officials of the Tribal Court indicate that new boundary agreements are not forthcoming and land disputes are on the rise.[25] The uncertainty of land boundaries helps explain why there are so few consolidations and transfers outside the family. In addition, Tribal Council regulations allow permit subleasing *only* among family members and forbid it among outsiders.[26] Such restrictions result in a reduction of wealth for the Navajo, a people who by any standard are already poor. The predominance of small herds, then, necessitates other sources of income, such as wages, welfare payments, and federal emergency feed grain programs. Since 1958, at least eight such programs have been necessary to supplement the grass produced on the overstocked range and to avoid animal starvation. Grain distribution has varied from 115 million pounds of milo in 1956 to nearly 82 million in 1972 to over 6 million in 1978.[27]

The long-run efficiency effects of the Indian Service's failure to precisely define land rights in 1941 is reflected in the rising number

NAVAJO TRIBE GRAZING FORM NO. 2 9-58-10M

THE NAVAJO TRIBE

WINDOW ROCK, ARIZONA

____
DISTRICT

GRAZING PERMIT
BILL OF SALE AND TRANSFER AGREEMENT

THIS IS TO CERTIFY that in consideration of ________________________ Dollars ($__________)

Receipt of which is hereby acknowledged, I ____________________ (NAME OF GRANTOR) __________ (CENSUS NO.)

do hereby grant, sell, transfer, assign and deliver to ____________________ (NAME OF GRANTEE) __________ (CENSUS NO.)

__________ (NUMBER) sheep units, including __________ (NUMBER) horses from my current grazing permit to be grazing in the District Named above.

GRAZING PERMIT DATA BEFORE TRANSFER

NAME	CENSUS NO.	PERMIT NO.	DATE ISSUED	NO. SHEEP UNITS	NO. HORSES
Grantor:					
Grantee:					

GRAZING PERMIT DATA FOR NEW PERMITS

NAME	CENSUS NO.	BRAND	NO. SHEEP UNITS	NO. HORSES	SEASON
Grantor:					
Grantee:					

WITNESS TO GRANTOR'S SIGNATURE DATE — GRANTOR

WITNESS TO SPOUSE'S SIGNATURE — SPOUSE OF GRANTOR

WITNESS TO GRANTEE'S SIGNATURE DATE — GRANTEE

Recommendation:

CHAIRMAN, DISTRICT GRAZING COMMITTEE DATE — Concur — BRANCH OF CREDIT

Approved: ____________________
SUB-AGENCY SUPERINTENDENT DATE

GRANTEE

Fig. 4. Bill of sale and transfer agreement for grazing permit.

of land disputes and the limited number of land and sheep transfers and consolidations. In addition, the policy of assigning rights to small herders rather than large ones meant that subsequent transaction costs would be large.[28]

In the early 1940s, the Tribal Council repeatedly endorsed the Indian Service's permit policy. A resolution in 1940 called for the distribution of grazing permits (which had yet to be assigned) "from large to small owners."[29] Two resolutions in 1941 and 1942 called for a maximum limit of 350 sheep units per individual.

> Whereas the large owners had always and continuously dominated control of the range; and Whereas, the small owners never had any chance to build up or increase their herds. . . . Therefore, be it resolved, it is now an opportune time, under the existing special grazing permit of 350 sheep units, to help this class of people.[30]

In 1943, the Council called for *differential enforcement of grazing regulations,* which allowed small herders to exceed their permitted number.[31] Later in that year, the Council demanded the abolishment of the range demonstration areas that had been fenced by the Indian Service because such areas were "used by a few privileged Navajos."[32] Paradoxically, the Council recognized that those areas had the only good range to be found.

Throughout the 1940s, the Indian Service appears to have strictly enforced its grazing regulations under the authority of a 1940 federal court ruling.[33] This was done independently of the tribe; but with the lapse of the stock reduction program in 1948, federal policy began to promote greater tribal authority in grazing matters.

Property Rights and the Post-Stock Reduction Period

From 1948 to 1956, the BIA continually tried to transfer responsibility for grazing regulation to the Tribal Council. The record indicates that the Council did not want to assume that responsibility. Young shows that the Council repeatedly requested a postponement of the deadline for implementing its rules.[34] In 1953, the Council did create grazing committees for each district to enforce the regulations and handle disputes.[35] In 1956, the BIA withdrew from active enforcement, assuming a posture of an advisory agency. Since that year, the tribe has largely relinquished its responsibility by relying on voluntary compliance with the authorized permit size and grazing regulations.

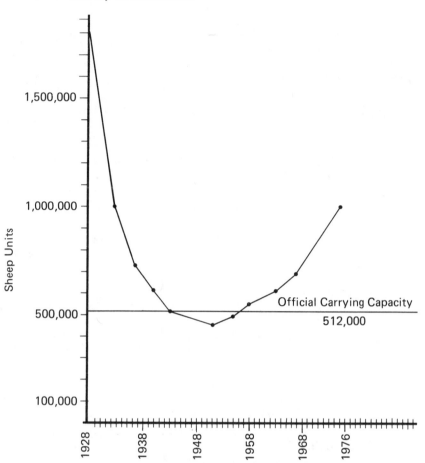

Fig. 5. Animal units on the Navajo reservation and carrying capacity, 1928–76. (Data for 1928–59 from Young, *Navajo Yearbook;* for 1943, Indian Service study; for 1960–76, tribal summer counts.)

This policy has failed; the number of animal units on the range has rebounded rapidly, approaching in 1976 the numbers that existed prior to the stock reduction (see fig. 5). Overgrazing has also returned.

It is possible that conditions are worse than indicated. While 512,000 units remain the official carrying capacity figure used by the tribe to calculate overstocking, there is some evidence that carrying capacity has fallen over time.[36] The tribe's grazing regulations, adopted in 1956, contain few specifics on grazing rights or their enforcement. While the district grazing committees are charged with the arbitration of land disputes, they tend to pass that responsibility to tribal courts, because enforcement of the regulations is politically unpopular.[37] Land disputes are on the rise, and the courts are backlogged.[38] A reading of court opinions reveals two clear points: first, there have been significant delays in resolving trespass conflicts (some have been in progress for ten years), and second, the grazing committees have failed to act.[39] The resulting uncertainty over land boundaries not only inhibits the necessary consolidation of herds (permits), it encourages overgrazing, particularly in the contested areas where rights are both disputed and dispersed.

Implications and Empirical Tests

When regulatory power was transferred from the BIA to the tribe, the reservation as a whole was not overgrazed and stocking was at the carrying capacity (see table 4). Because land boundaries become less clear as permits are divided, a testable implication is that overgrazing should be more common where there are smaller herds. Data confirm the hypothesis that overgrazing (the ratio of actual stocking to permitted stocking) is higher for the smaller permit holders than it is for the larger ones (see table 4). In addition, using 1957 cross section data on percent overgrazing and the percent of permit holders having fewer than 100 sheep units for fifteen reservation districts yielded a correlation coefficient of 0.61, which is significant at a level of one percent.[40] There is, then, a strong relationship between overgrazing and permit size. This is not only due to the increased vagueness of land boundaries as the customary use areas are divided, but also because of the apparent reluctance of the tribe to enforce grazing regulations on smaller operators.

The correlation between overgrazing and permit size can be used to predict the effect of further declines in the number of sheep in each herd. Should the number of permit holders increase while the carrying capacity of the range and the permitted number of animal units remain constant, then changes in the degree of overgrazing should be strongly associated with the districts with the greatest increase in the number of permit holders. From 1957 to 1967, the number of permit holders grew by almost 18 percent, mainly through inheritance, with no corresponding increase in carrying capacity or permitted numbers of sheep units. In spite of that, total stocking

TABLE 4. Analysis of Grazing Permits and Livestock Numbers Operated by Permittees, 1957

Size of Herds (in sheep units)	Number of Permittees	Permitted Number of Sheep Units	Actual Number of Sheep Units Grazed	Percent of Permitted Number
1–25	2,194	34,397	44,174	128
% of total	27.2	6.4	8.6	
Cumulative %	27.2	6.4	8.6	
26–50	1,900	72,449	76,661	106
% of total	23.5	13.4	14.9	
Cumulative %	50.7	19.8	23.5	
51–75	1,460	91,972	94,545	103
% of total	18.1	17.0	18.4	
Cumulative %	68.8	36.8	41.9	
76–100	896	78,190	73,708	94
% of total	11.1	14.5	14.3	
Cumulative %	79.9	51.3	56.2	
101–150	858	103,475	91,026	88
% of total	10.6	19.2	17.7	
Cumulative %	90.5	70.5	73.9	
151–200	429	72,583	64,245	89
% of total	5.3	13.4	12.5	
Cumulative %	95.8	83.9	86.4	
201–50	178	40,067	32,070	80
% of total	2.2	7.4	6.2	
Cumulative %	98.0	91.3	92.6	
251–300	114	31,171	23,425	75
% of total	1.4	5.8	4.6	
Cumulative %	99.4	97.1	97.2	
301–50	48	15,918	14,484	91
% of total	.6	2.9	2.8	
Cumulative %	100.0	100.0	100.0	
Total	8,077	540,222[a]	514,338	
Average				95

The 1943 carrying capacity is 512,922 sheep units yearlong; the ratio of carrying capacity to use equals 1.002.

Source: 1957 BIA Range Study.

Note: The permitted number of sheep units exceeds the carrying capacity due to the issuing of special permits.

a. Includes supplemental permits in effect and seasonal permits actually grew, increasing from 514,338 sheep units in 1957 to 605,773 in 1967.[41] Comparing the change in the percent of overgrazing and the percent change in the number of permit holders for each of the fifteen districts yielded a correlation coefficient of 0.68. Again, the hypothesis is supported by the evidence: the districts with the greatest increase in small herds also experienced the greatest increase in overgrazing.

Extending the test for nine more years reveals a similar pattern. From 1967 to 1976, total stocking on the reservation rose by 67 percent, to nearly a million sheep units, a population that corresponds to the pre-stock reduction period of the early 1930s.[42] While the data are imperfect, the relationship between the percent of overgrazing and the percent of permit holders with less than 100 sheep units in each of the districts shows a correlation coefficient of 0.77 for cross section data in 1976.[43]

Fencing, Tribal Policy, and Property Rights

Since overgrazing has actually increased since the late 1950s, it seems clear that Navajo herders have an incentive to establish a more definite set of property rights over the range. The literature on institutional change has argued that property rights become more exactly defined and enforced when "it becomes economic for those affected by externalities to internalize benefits and costs."[44] In this section, we consider the costs of defining customary use area boundaries and fencing.

Fencing (particularly the decline in the cost of barbed wire) has been acknowledged as a major factor in fostering the establishment of property rights over western range land.[45] Today, much of western range land is fenced or otherwise controlled by natural barriers. A notable exception is the Navajo reservation, where only 3 percent of the range land is enclosed.[46] The limited amount of fencing on the reservation is even more of an anomaly since the federal government and the tribe have, since the late 1960s, agreed to bear material and installation costs.

Fencing is only briefly discussed in the *Tribal Grazing Handbook*, reflecting, perhaps, the earlier bias against it. The Tribal Council requires that before fencing can occur, the individual must have *unanimous* approval of all neighbors who may be affected by it.[47] Further approval is necessary from the district grazing committee, the chapter president and vice president (the chapter is the smallest governmental unit on the reservation), the individual's Tribal Council representative, the Tribal Resources Committee, and the Navajo Agency superintendent (BIA). Discussions with BIA and tribal range officials indicate that the unanimity rule poses the greatest barrier to fencing. Buchanan and Tullock point out that, despite their attractiveness for equity reasons, unanimity rules impose prohibitive transaction costs, necessitating the adoption of alternative decision rules, such as a majority vote.[48]

Although numerous Navajo have attempted to secure agreements from herdsmen surrounding their areas, most have failed. BIA officials in the western part of the reservation indicated that the success rate was about one out of five.[49] Even for those who do succeed, the time lapse between initiation and obtainment of unanimity typically involves three or more years.

The frequency of successful delineation of boundaries can, of course, be stated in the traditional framework of a benefit-cost analysis. Assuming that transaction costs in establishing an agreeable boundary are relatively constant per foot of boundary line, it follows that agreement should be more prevalent the higher the value of the land, *given enforcement*. For example, if a certain type of land is more responsive to control management and improvement practice once fencing is completed, we expect the frequency of agreement over boundaries to be higher than for areas where range improvement practices require a greater investment. In addition, larger herders should have more incentive to fence than smaller herders. There are three reasons for this: First, simple geometry implies that the periphery of an area, such as a circle or square, does not increase in constant proportion to the area. Thus, transaction costs per acre fall with an increase in area. Second, as permits are broken up, the customary use area boundaries become increasingly vague and, accordingly, small permit holders are less likely to fence. This, of course, would be contrary to the assumption that transaction costs involved in negotiation are uniform per foot of boundary line. The predicted sign, however, is still unambiguous; the more sheep units per permit, the greater the probability of agreement as to boundaries.[50] Third, with variable precipitation due to thundershowers that leave surrounding areas dry, the enclosure of small areas greatly restricts mobility and increases the risk of poor forage for small herds.[51] Thus, the larger the area, the lower the variance in annual rainfall and, assuming risk aversion, the higher will be the willingness to pay the transaction costs involved in fencing negotiations.

To test the relation between permit size and fencing agreements, we examined information on each area fenced. Data were obtained from files on land management units available in BIA offices. It is common that a fenced area generally encloses more than one permit holder. In part, this reflects the tendency of a family group to consolidate their herds and permit holdings and then to negotiate with those in the surrounding area. There are notable exceptions, however, as perimeter fencing is frequently followed by cross fencing of the initial area.

Our analysis concentrated on the initial perimeter fencing agreements. As the land within the fenced area generally contains more than one permit holder, we avoided the potential bias of combining them into a single unit. Rather, we compared the average permit size within the fenced area with the average permit size for the entire reservation at the time the fence was approved. Of fifty-five fenced areas, forty-two had an average permit size *greater* than that for the entire reservation. Sign test utilization (the distribution of the values cannot be assumed normal) shows this result to be significantly positive at a level of one percent.[52] It is worth noting that over one-fourth of the permits were twice as great as the average.

Generally speaking, most of the rangeland on the Navajo reservation could benefit from enclosure by fencing. For our purpose, however, it is important to be able to distinguish which types of land are likely to benefit the most. In most cases, the land that is likely to respond best to range management techniques, such as reseeding of overgrazed areas, is located in the higher elevations where rainfall is more abundant. BIA officials indicate that successful reseeding normally requires at least ten inches of rainfall per year. For the western and middle sections of the reservation, the required rainfall occurs most commonly above 6,500 feet. The areas receiving less than ten inches of rainfall do not respond well to reseeding and require longer periods of very limited use if regeneration is to be successful. Thus, returns to fencing are greater where rainfall is more abundant. Data reveal that 81 percent of the fenced areas in the western and middle section of the reservation were above 6,500 feet with the majority being above 7,500 feet. Since most of the range land is below 6,500 feet, this confirms the hypothesis that agreement on boundaries will increase where the value of land, given enforcement, is higher.

The data for Shiprock Agency, which covers most of the northeastern part of the reservation, do not appear to confirm the hypothesis. Only eight out of fifteen fenced areas average ten inches or more rainfall per year. However, six of the areas with less than ten inches of rainfall per year are located in District 13. The district comprises only 14 percent of the land area of the agency, but contains nine out of the fifteen fenced areas. District 13 has the highest average permit size for the entire reservation and historically has never exceeded the estimated carrying capacity of the range. Both permit size and rainfall contribute positively to fencing agreements, the former apparently being the dominant factor in District 13.

Conclusion

In 1975, the United States Commission on Civil Rights reported a widening gap between the average American and Navajo per capita income from 1950 to 1972. In 1972, Navajo income was approximately one-quarter the national average.[53] It is clear that the Tribal Council and the BIA have contributed to this condition by failing to adopt policies that would minimize transaction costs and thus maximize wealth from land resources. A preoccupation with equity has resulted in the progressive shrinking of individual herd size and severe restrictions on fencing. In addition, the splitting of sheep permits leads to confusion about grazing boundaries and fewer incentives to fence (particularly as individual plots become smaller and smaller). The Navajo range has all the characteristics and problems of a common property resource.[54] The very existence of the numerous regulations regarding grazing and fencing underscores the lack of control that individual herders have over the intensity of land use. The problem of overgrazing is not a lack of regulations, but an unwillingness to enforce the existing ones. Since the BIA has become an advisory agency, enforcement rests with the Tribal Council. The policy of granting only usufruct rights not only encourages overgrazing (land that is not grazed is lost to another), it gives responsibility for resource use to the Tribal Council rather than to those who actually occupy the land.

Current policy appears designed to maximize votes by allowing almost everyone to have some sheep at the cost of severe wind and water erosion and deterioration of the range. This program is subsidized by federal emergency feed grain shipments, welfare payments, and wage work.[55] Such political conditions make it difficult for the Navajo to consolidate herds, to fence, and to make a living from raising sheep. As economic theory predicts, the lack of precise land boundaries and the failure to enforce grazing regulations explain chronic overgrazing on the Navajo reservation.

Notes

1. For discussions of early grazing practices on the Navajo reservation, see Donald L. Parman, *The Navajo and the New Deal* (New Haven: Yale University Press, 1976), p. 10; and Lee Muck, "Survey of the Range Resources and Livestock Economy of the Navajo Indian Reservation" (Washington, D.C.: U.S. Department of Interior, Office of Land Utilization, 1948), p. 6. We recognize that the definition of overgrazing is based primarily on biological factors and may not be consistent with some steady state solution derived using both biological and economic criteria. However, the estimated carrying capacity of the range will be used in this paper as a datum from which to measure changes in resource use as a consequence of institutional constraints.

2. Discussions of the importance of sheep and grazing in Navajo life can be found in Robert W. Young, *Navajo Yearbook* (Window Rock, Ariz., 1961), pp. 143–51; David F. Aberle, *The Peyote Religion Among the Navajo* (Chicago: Aldine-Atherton, 1966), pp. 23–52; Phelps Stokes Fund, *The Navajo Indian Problem* (New York: Phelps Stokes Fund, 1939), pp. 1–9; and Ruth Underhill, *The Navajos* (Norman: University of Oklahoma Press, 1956), pp. 236–53. Most of these writers attribute the nomadic and pastoral lifestyle of the Navajo to cultural values, but they do not examine the property ownership arrangements. For example, in the study for the Phelps Stokes Fund, we find the following: "Basically, then, the Navajo problem on its economic side is the failure to reconcile the tenacious and anxious devotion of the Navajo people to their customs and their language with the aggressive determination of able and devoted soil conservationists to save the soil elementally necessary to the very existence of the Navajos" (p. 3).

3. See Harold Demsetz, "Toward a Theory of Property Rights," *American Economic Review* 57 (May 1967):350; Armen Alchain and Harold Demsetz, "Property Rights Paradigm," *Journal of Economic History* 33 (March 1973):24; Steven N. S. Cheung, "The Structure of a Contract and the Theory of a Non-Exclusive Resource," *Journal of Law and Economics* 13 (April 1970):49–70; and Lance E. Davis and Douglass North, *Institutional Change and American Economic Growth* (Cambridge: Cambridge University Press, 1971).

4. For discussions of Navajo property arrangements, see Felix S. Cohen, *Handbook of Federal Indian Law, 1907–1953* (Albuquerque: University of New Mexico Press, 1971); and Van Valkenburgh, "Navajo Common Law," *Museum Notes* (Museum of Northern Arizona, Flagstaff) 9, no. 4 (1936):20–22.

5. For a description of conditions leading to the formation of the Tribal Council, see Lawrence C. Kelly, *The Navajo Indian and Federal Indian Policy* (Tempe: University of Arizona Press, 1968), pp. 49–69.

6. Population figures are from Denis F. Johnston, "An Analysis of Sources of Information on the Population of the Navajo," *Bureau of American Ethnology Bulletin* 192 (Washington, D.C.: Government Printing Office, 1966), p. 87; and the reclamation discussion is from Paul W. Gates, *History of Public*

Land Law Development (Washington, D.C.: Government Printing Office, 1968), pp. 619–89.

7. For an outline of conditions leading to the 1933 stock reduction, see Aberle, *The Peyote Religion*, pp. 52–55; and Muck, "Survey of Range Resources."

8. The stock reduction is described by Aberle, *The Peyote Religion*, pp. 52–79; Muck, "Survey of Range Resources"; Kelly, *The Navajo Indians*, pp. 155–67; and Parman, *The Navajo and the New Deal*, pp. 33–293. In these and other studies, there is emphasis on the mistrust that developed between the Navajo and the Indian Service. While this was in part likely due to political conditions on the reservation, Collier was also responsible. He promised the tribe that the New Mexico boundary would be extended if the reduction were carried out. The Senate subsequently killed the bill. Sheep prices in the text are from U.S. Department of Agriculture, Arizona Crop and Livestock Reporting Service, *1964 Arizona Agricultural Statistics*, p. 110; and Aberle, *The Peyote Religion*, pp. 58–59.

9. A sheep unit is the grazing of one sheep for one year with sheep and goats equaling one unit each and cattle and horses equaling four and five units. Only mature animals are included, so the stocking count is an underestimate of the true number of animals on the range.

10. Young, *Navajo Yearbook*, 1961, pp. 167–68.

11. Parman, *The Navajo and the New Deal*, p. 45; and U.S. Senate, Committee on Indian Affairs, *Survey of Conditions of the Indians in the United States*, *Pt. 34* (Washington, D.C.: Government Printing Office, 1937).

12. Parman, *The Navajo and the New Deal*, p. 49.

13. Aberle, *The Peyote Religion*, p. 55.

14. In the initial stock reduction, the large herders resisted and were able to win a 10 percent across-the-board cut for all herds. Yet, by the 1940s, the brunt of the program shifted to them. For further support of this statement, see Aubrey W. Williams, Jr., *Navajo Political Process* (Washington, D.C.: Smithsonian Institute Press, 1970), p. 28.

15. For example, see National Association of Indian Affairs, *The Navajo and the Land* 26 (1937):7; Phelps Stokes Fund, *The Navajo Indian Problem*, p. 11; and Mary Shepardson, "Navajo Ways in Government," *American Anthropological Association* 65 (June 1963):67. In contrast, Downs seems to recognize the efficiency effects. See James F. Downs, *The Navajo* (New York: Holt, Rinehart & Winston, 1972), p. 132.

16. This situation is similar to McCloskey's finding that most of the literature on English common lands and enclosures in the seventeenth and eighteenth centuries emphasizes the distributional questions and neglects the output impacts. See Donald N. McCloskey, "The Persistence of English Common Fields," in *European Peasants and Their Markets*, ed. William N. Parker and Eric L. Jones (Princeton: Princeton University Press, 1975), pp. 88–113.

17. Parman, *The Navajo and the New Deal,* p. 56, and Kelly, *The Navajo Indian,* p. 160.

18. Navajo Tribal Council, *Resolutions of the Navajo Tribal Council 1934* (Window Rock, Ariz., 1934), p. 258.

19. Arizona Crop and Livestock Reporting Service, U.S. Department of Agriculture, *1966 Arizona Agricultural Statistics,* p. 110.

20. Parman, *The Navajo and the New Deal,* p. 183.

21. Muck, "Survey of Range Resources," exhibit F.

22. Aberle, *The Peyote Religion,* p. 67; and Muck, "Survey of Range Resources," pp. 11, 12. The Muck report includes the federal regulations for assigning permits and regulating grazing.

23. U.S. Department of Agriculture, New Mexico Crop and Livestock Reporting Service. Also see N. B. Pingrey, "Marketing Western Range Sheep and Lambs," *Agricultural Experiment Station Bulletin* (Albuquerque) 434 (April 1959). Pingrey indicates that 50 percent of the sheep ranges in the western states fall within the 350 to 3,499 sheep unit size clause.

24. Young, *Navajo Yearbook,* 1969, p. 165.

25. See Navajo Nation, Judicial Branch, "Annual Report," 1972-78.

26. See Navajo Tribal Council, *Navajo Reservation Grazing Handbook* (Window Rock, Ariz., 1962), pp. 27-28.

27. Federal emergency feed programs were as follows:

1956	115,000,000 lbs. milo @ 2.29 per 100 = $2,633,500
1958	12,000,000 lbs. milo @ 2.23 per 100 = 267,600
1959	24,000,000 lbs. milo @ 1.98 per 100 = 475,200
1961	16,800,000 lbs. milo @ 1.87 per 100 = 314,160
1971-72	81,603,192 lbs. milo @ 3.81 per 100 = 3,109,082
1972-73	40,676,575 lbs. milo @ 4.95 per 100 = 2,013,492
1974-75	40,057,812 lbs. oats or
	1,251,812 bu. oats @ 1.68 per 100 = 2,103,044
1978	6,400,000 lbs. milo @ 4.89 per 100 = 3,129,600

Nominal prices are used. Rainfall is an important variable in explaining fluctuations in the program; 1956 was a particularly severe drought year. Feed grain statistics are from the Bureau of Land Operations, Bureau of Indian Affairs, Window Rock, Ariz.; prices are from U.S. Department of Agriculture, Agricultural Statistics, *Annual Report,* 1961, 1977.

28. Harold Demsetz, "Some Aspects of Property Rights," *Journal of Law and Economics* 9 (October 1966):66.

29. Navajo Tribal Council, *Tribal Council Resolutions,* June 1940, p. 228.

30. Navajo Tribal Council, *Tribal Council Resolutions,* June 1942, p. 234.

31. Navajo Tribal Council, *Tribal Council Resolutions,* July 1943, p. 235.

32. Ibid., p. 172.

33. Young, *Navajo Yearbook,* 1961, p. 155.

34. Ibid., p. 155–56.

35. Williams, *Navajo Political Process,* pp. 30–33.

36. For a discussion of recent range conditions, see Aberle, *The Peyote Religion,* p. 83.

37. See Williams, *Navajo Political Process,* pp. 32–33, for a discussion of why grazing committees fail to enforce the regulations.

38. Interviews with Tribal Court officials and personnel at the tribal and BIA range offices confirm that disputes are on the rise. Unfortunately, data on the number of cases handled by the grazing committees over time are not yet available.

39. For example, in *David Brewster* v. *John Bee and Harry Benally* (Navajo Court of Appeals, March 3, 1977), the court repeatedly emphasized both the time lags in processing cases and the failure of the grazing committee to act.

40. Fifteen rather than eighteen districts were used for the following reasons: District 6 is the Hopi reservation and, therefore, outside our analysis, and Districts 15 and 16 lie in part off the reservation and involve special problems that make them dissimilar from other districts.

41. 1957 stocking, BIA, and 1967 tribal summer counts. See U.S. Department of Interior, Bureau of Indian Affairs, *Roundup of Navajo Grazing Permits and Livestock Inventory 1957* (Window Rock, Ariz., 1957).

42. 1976 tribal summer count.

43. Separate data for Districts 7, 14, 17, and 18 were not available. Total agency figures were used for these districts. For time series data, while the number of sheep increased dramatically from 1967 to 1976, the rate of increase in the number of permits was only 5 percent. This indicates that, at least in the recent period, the decline in individual herd size from splitting permits is not the only factor explaining the continued rise in overgrazing. Another possible explanation is the federal emergency feed grain program.

44. Demsetz, "Theory of Property Rights," p. 354; Davis and North, *Institutional Change;* and Gary D. Libecap, "Economic Variables and the Development of the Law: The Case of Western Mineral Rights," *Journal of Economic History* 38 (June 1978): 338.

45. Terry L. Anderson and P. J. Hill, "The Evolution of Property Rights: A Study of the American West," *Journal of Law and Economics* 18 (April 1975):163–79.

46. Total fenced acres on the reservation:

Shiprock agency	111,789
Fort Defiance agency	118,445
Chinle agency	73,907
Tuba City agency	163,600
Total	467,741

The total figure is approximately 3 percent of the total area of the reservation.

47. *Grazing Handbook,* 1962.

48. See James Buchanan and Gordon Tullock, *The Calculus of Consent* (Ann Arbor: University of Michigan Press, 1965), pp. 85–96; and Mancur Olson *The Logic of Collective Action* (Cambridge: Harvard University Press, 1965). In a similar problem, McCloskey discusses the impact of unanimity rules in preventing voluntary land enclosures in seventeenth and eighteenth century England. This led to the adoption of Parliamentary statutes to force enclosures. See Donald N. McCloskey, "The Economics of Enclosure: A Market Analysis," in *European Peasants and Their Markets,* pp. 127–36.

49. Interviews at the Tuba City and Chinle BIA agencies, June 1978.

50. The third reason has also been discussed by McCloskey in regard to scattered plots in preindustrial Britain; when land quality is variable, the need to reduce risk increases. See McCloskey, "The Persistence of English Common Fields," pp. 113–16.

51. Discussions with weather bureau officials in Albuquerque indicate that summer rainfall amounts vary significantly in the Southwest for sites at the same elevation that are in excess of one mile from one another. This means that reasonably close locations are likely to receive different percipitation amounts from year to year. For instance, in 1976, the Clovis reporting station had nearly twice as much rainfall as a neighboring station at the same elevation thirteen miles away.

52. See W. J. Dixon and F. J. Massey, *Introduction to Statistical Analysis,* 3d ed. (New York: McGraw-Hill, 1969), pp. 335–37.

53. U.S. Commission on Civil Rights, *The Navajo Nation: An American Colony* (Washington, D.C.: Government Printing Office, 1975), p. 44.

54. Garrett Hardin, "The Tragedy of the Commons," *Science* 162 (1968): 1243–48. Census data from 1970 show that wage earnings, welfare payments, and social security are the major sources of income for more Navajo families than grazing. See Navajo Research and Statistics Center (Window Rock, Ariz., 1978).

55. Young, *Navajo Yearbook,* 1961, pp. 213–29.

Dams and Disasters: The Social Problems of Water Development Policies

by Bernard Shanks

Early in President Carter's administration, an optimistic attempt was made to cut off funding for approximately 30 dams and water development projects. A leading environmentalist, in a euphoric mood, stated, "The President is in the process of destroying the cozy relationship among powerful governmental bureaucracies, pork-barrel Congressional committees and water development and user interests."[1] His elation was short-lived. Within a year most of the "hit list" was restored. Projects authorized by Congress continued to be constructed despite social and environmental costs. A backlog of 828 water projects with a $34 billion price tag remained authorized despite efforts to reexamine project rationale and cancel inefficient projects. Water development agencies—the Army Corps of Engineers, the Bureau of Reclamation, and the Tennessee Valley Authority—and congressional public works committees and subcommittees remain as cozy with their clients as ever. President Carter remains one of a long line of individuals who have tilted the water development windmill. Not only presidents and environmentalists, but those concerned with efficient government have broken their lances against the triumvirate that guards the water resource development territory.

It is almost a truism that federal water projects have resulted in substantial environmental problems. It is also obvious that water developments have brought substantial economic benefits to some groups. This paper focuses on the process of development, its problems, and the allocation of benefits and costs over time, space, and a pluralistic society. While it only summarizes the abundant literature, this paper hopes to explain some of the reasons for the failure of Carter's "hit list" and to offer a few suggestions.

Historic Background

Water developments are the oldest form of federal environmental programs. Over 150 years of development process and programs are available for evaluation. In addition to being the oldest type of national resource development, water projects were the first federal projects to have a specific economic test required during planning. The requirement that benefits exceed costs resulted in the complex and controversial field of cost-benefit analysis. Water resources have spawned three federal agencies concerned with major development projects and another, the Soil Conservation Service, concerned with the construction, planning, and assistance of smaller scale projects. The agencies are complex and have a variety of constituents and advocates.

Water developments are numerous and popular. By 1977, nearly 50,000 dams twenty-five feet or higher had been constructed in the United States.[2] Many of the large dams are private, although the largest are federal. Federal financial aid and technical assistance in planning and construction have subsidized many of the 2.5 million small dams less than twenty-five feet high or with a capacity of less than fifty acre feet of water.[3] Federal water policy is a major geomorphic process in the United States.

The importance of water development as a policy program might be illustrated by contrasting it with the federal program for stream conservation. In 1968, Congress established a policy of protecting some streams in their free-flowing condition by passing the National Wild and Scenic Rivers Act (82 Stat. 906). Initially protecting several rivers from dams, the act provided for additional rivers to be added to the national system. After ten years, some 1,700 miles of free-flowing rivers were under some form of federal protection.[4] This mileage represents 1 mile for each major dam (over twenty-five feet high) constructed in the United States annually. Another way to contrast federally supported water development projects with river construction is to note that for every thirty-three major dams (over twenty-five feet high) in place there is 1 mile of free-flowing river protected by the Wild and Scenic Rivers Act. If all dams, large and small, are compared to free-flowing streams, the contrast is even more dramatic.

For every dam in the United States there remains approximately 1.1 meters (3.6 feet) of wild and free-flowing streams.

Federal Water Development Agencies

Federal water development policy began with a 1824 Supreme Court decision that gave the federal government the power to regulate interstate commerce, including navigation on rivers and harbors.[5] The decision cleared the way for the federal government to benefit trade by developing canals and river improvement projects. Lacking a large bureaucracy, the government assigned the Army Corps of Engineers to carry on the modest projects of removing snags, rocks, and sandbars that interfered with barge and steamboat traffic. From an unpretentious beginning, the federal mission grew increasingly complicated. Successive acts of Congress gave the Corps of Engineers increased responsibility to protect the navigability of rivers and harbors. By the end of the nineteenth century, they had responsibility for bridges, wharves, piers, channels, harbors, diversion of water, and the disposal of refuse. Located in the Department of Defense, the Army Corps of Engineers has been a long-standing organizational anomaly among natural resource programs. Its employees are largely civilian, and their mission is far removed from national defense.

Flood control on a federal level began in 1879 with the establishment of the Mississippi River Commission. The 1902 Reclamation Act (32 Stat. 388–90) established a new agency, the Bureau of Reclamation, for the purpose of supplying water for irrigation. Its role was limited to the arid western states. Originally the bureau was to be self-supporting: it would loan money for irrigation projects, and the fees from the sale of water would repay the cost of construction plus interest. Soon the bureau expanded its mission to include power, flood control, recreation, and fisheries. This permitted a portion of nonirrigation benefits to be charged against the cost of construction and operation. But the sale of water for irrigation could not cover the bureau's expenses, and the notion that the bureau would be self-supporting was abandoned.

The most recent expansion of the federal government's role in water development came with Public Law 566, passed in 1954. This legislation gave the Soil Conservation Service (SCS) in the Department of Agriculture the responsibility for watershed projects and dams with less than 25,000 acre feet of capacity. It provided technical assistance and loans to private parties for smaller dam projects. Restricted in project size, the SCS was a fourth federal agency in a fourth department to become involved with water construction projects.

The third and largest shift in policy came when the Tennessee Valley Authority (TVA) was established in 1933. The TVA rose out of the Depression as a unique regional development corporation with comprehensive program responsibility for both water and related land resources. Its independent authority exceeds any other natural resource agency. President Roosevelt advocated several similar authorities for the remaining watersheds. How the Corps of Engineers and Bureau of Reclamation responded to these proposals is an interesting study in bureaucratic reactions to threatening policies. It also illustrates the problem of developing institutional alternatives to the present water development situation.

A Case Study

In the spring of 1943, the Missouri River flooded adjoining lands to heights not reached since 1881. Fifty million dollars in damages resulted, including damage to 2.4 million acres of farm land. Congress requested that the Corps of Engineers review the flood control needs of the Missouri. The Army Corps of Engineers referred to their 1,245-page Missouri River report of 1934 and, by the spring of 1944, had compiled a 12-page "Pick Plan" for flood control and navigation on the lower Missouri Basin. The major projects, in addition to levees, were five large main stem dams and six dams on major tributaries. The plan promised to stabilize the valley's economic life, encourage industry and civic growth, and relieve the distress caused by the river.[6] The Pick Plan was criticized because it ignored upstream irrigation interests, and proposed developments were not comprehensive.

In contrast, the Bureau of Reclamation spent five years preparing the more comprehensive and detailed Sloan Plan. While the plan stressed irrigation (reflecting the influence of John Wesley Powell), it also included flood control and navigation on the lower Missouri. It proposed ninety reservoirs, facilities to irrigate 4.7 million acres of land, and projects to produce 4 billion kilowatts of water generated power.

During the summer of 1944, Congress was faced with two agencies proposing two different plans for the same watershed. As might be expected, each agency publicized its own plan and criticized the other's. Newspapers ran editorials extolling the virtues of one plan or the other. Congressmen advocated whichever plan would most benefit their constituencies. The controversy heightened until on September 21, 1944, President Roosevelt asked Congress for the creation of the Missouri Valley Authority, to be patterned after the TVA. Unable to control his two major water agencies, Roosevelt proposed a new one.

On October 16 and 17, representatives of the Corps of Engineers and the Bureau of Reclamation met in Omaha and reached a one-page agreement known as the Pick-Sloan Plan. In two days, the entire development plan for the largest river in North America was decided by two agencies without any outside involvement and in opposition to a presidential proposal. In a classic description of the Pick-Sloan Plan, one paper editorialized:

> The resumption of this old and apparently irreconcilable feud between two vested government interests convinced many people that the time had come to cut the Gordian knot by advancing the MVA (Missouri Valley Authority) idea. The idea would reserve the Missouri Valley from contending factions and place it under harmonious and scientific, but above all, under united and non-political management. As the MVA idea took instant hold upon the imagination of the country, and won the ultimate endorsement of the President in a special message to Congress, a strange and wonderous thing occurred. The feudists, fearful of the MVA idea, lest it invade their bureaucratic precincts, began to murmur softly to each other and how marvelous they relate—a marriage has been arranged. A loveless, shameless, shotgun marriage of convenience, arranged between old and bitter enemies, not only to kill off MVA, but to save the interest jealously guarded by two powerful government agencies.[7]

Unable to agree on the best projects, the Corps of Engineers and the Bureau of Reclamation decided to build everything. Reconciliation meant that each agency became reconciled to the works of the other. One Pick project was eliminated, but only because the site would be submerged after construction of a larger Sloan project downstream. Only days before, the bureau's regional director had testified that the Garrison Dam site was not only wasteful but dangerous. In the compromise, he approved the project. The Pick-Sloan Plan said nothing about allocation of water to navigation, flood control, power, and other uses. Important questions of land use, regional distribution of benefits, and comprehensive resource development were not asked. Navigation won the lower basin, irrigation won the upper basin, and no one knew if enough water existed for both.

For over thirty years Congress provided $6 billion and the two agencies constructed dams. But no federal agency has examined the effectiveness of the two-day plan or even asked if its objectives were achieved. Only one thing is certain: nowhere in the history of water policy were so many dams approved in such a short time by so few people.

Water Development Evaluation

Economic analysis of project value began during the New Deal with the passage of the Flood Control Act of 1936 (49 Stat. 1570). In Section 1, Congress stated that "the Federal Government should improve or participate in the improvement of navigable waters or their tributaries including watersheds thereof, for flood control purposes if the benefits to whomsoever they may accrue are in excess of the estimated costs."[8]

Within a few years, benefit-cost analysis was used to justify all federal water development projects. Theoretically, the benefits had to exceed costs plus interest. It was soon recognized, however, that there were two broad categories of benefits and costs—tangible and intangible—and intangible benefits were not readily quantifiable. In addition, two types of tangible benefits and costs emerged: direct and indirect (or secondary). The secondary benefits permitted flexibility in the measurement of both benefits and costs. Analysts often enhance secondary benefits and underestimate secondary costs, particularly in human or social terms.

Rationale for Projects

The federal government's rationale for its involvement in water development has grown progressively more complex and confusing. Originally, the federal role was limited to navigation, but water-borne transportation on interstate waters has shrunk to a small part of the national transportation scene. Next, the government's role expanded to flood control, and today, alternatives such as land use planning and nonstructural alternatives for flood control are considered. The

Bureau of Reclamation's concern with irrigation has always been limited to the western United States, where farm surpluses and the diminishing returns from additional irrigation projects have been recent interests.

The low relative return of water resources allocated to agriculture, as opposed to other productive uses, has cut into the rationale for water irrigation projects. Private parties developed the best and most efficient sites early in the water development era and left the least productive sites to the federal government. Where individual projects could use the same equipment design and other factors of production, many large federal water projects were cheaper to construct because of economies of scale. The government also justified these projects by using low interest rates in benefit-cost analyses and allocating a portion of the costs to intangible recreation and other benefits. Today, the Office of Management and Budget requires higher interest rates, the project sites are poorer in quality, and alternative forms of development are often more efficient.

The rise in multiple purpose water projects complicated and duplicated the roles of the federal water development agencies. No longer is the Corps of Engineers only concerned with navigation and flood control. The Bureau of Reclamation is not restricted to the development of water resources for agriculture. All of the agencies are involved in power production and, in some cases, the transmission facilities as well. Fish, wildlife, and environmental programs are also important offshoots of federal programs. National concern over water quality has attracted water development agencies, who justify dam construction through "low flow augmentation" or the flushing of polluted streams during the dry season by water stored during the periods of high runoff. Recreation on many federal reservoirs is a major activity. Water projects also supply domestic and industrial water in some areas. The Bureau of Reclamation's role has expanded to include weather modification and complex efforts to increase snowfall and rainfall in the arid West.

The rationale for water development projects is easily as complex as the problems and concerns that have resulted from them. A constant problem has been that some water resources are public and some are private. While federal water development undoubtedly benefits both public and private parties, it is the allocation of costs and benefits that has been increasingly complex and controversial. Who actually receives the benefits of public projects? Who pays the costs?

Problems of Benefit-Cost Allocation

Several problems developed from federal water projects. First, the dams (approximately 1,000) were built on sites economically unsuited for private development, and economies of scale in design forced the developments to become larger and larger. At the same time, the problems associated with them grew. In the western United States, many of the best agriculture, wildlife, waterfowl, and other natural resources were located in the riparian zone along the major rivers. In many cases, the ranching economy was tied to river bottomland. Although the lands were sometimes a fraction of an economic unit, they were often the critical lands for livestock operations.

From the 1930s to the 1950s, water development agencies planned and constructed many massive federal water projects. At the same time, the technology of transferring power long distances was rapidly developing. Power and water resources became more mobile than human resources. Urbanization was proceeding, and many federal projects promised to reverse the migration from rural areas and to develop local economies. This did not happen. Power, industries, and jobs continued to flow to urban areas, and after the initial construction, many rural areas found themselves without new businesses. In many cases, these areas lost superior agricultural lands and part of the tax base to the reservoirs. The projects brought increased transportation problems, disrupted traditional patterns of communication and economic activity, and often brought in seasonal influxes of recreationists that were difficult and expensive to manage. Benefit-cost analysis and planning were at the core of the problem. Analysts had ignored the geographic distribution of both benefits and costs. Benefits of water projects flowed to the urban areas, but the costs, particularly noneconomic costs, often stayed close to the project site. Alternatives that would have enhanced local development were often not even considered.

Rural water storage or water rich areas tend to be exploited for those areas with the largest demand for and the political resources to achieve water development. For example, the loss of Missouri River irrigation water to barge traffic on the lower Missouri is questionable from an economic standpoint. The allocation of water from the water short areas in Montana to a water rich area along the lower Missouri increases existing economic and development disadvantages. Yet this was the basic pattern with the Fort Peck reservoir and dam, whose primary purpose was to increase water flows for barge traffic at the expense of lost tax revenues and agricultural land.

Did the Pick-Sloan Plan for the Missouri River meet its objectives? Decades after the completion of some projects, no significant differences could be found between developed and nondeveloped counties. The Plan failed to attract industry to the nonindustrial areas and reverse the emigration of people from rural areas. It neither reduced the numbers or percentages of people on welfare nor lowered the cost of public services. It did not even reduce flood damages.

The project replaced the subsistence economy of the Missouri River Indians, largely intact after thousands of years along the banks of the Missouri, with a welfare economy.[9] The Missouri River main stem dams often flooded the sparsely developed reservations and protected the land and cities downstream. The logic behind this was simple. The federal government could acquire Indian reservations at a fraction of the cost of urban areas. Flooding the reservations and protecting cities enhanced a positive cost-benefit ratio. But this approach, used not only on the Missouri but on many other rivers, brings special cultural problems not considered in most analyses.[10] Such projects can have a devastating effect on a minority or culture.

There were three important elements of the Indians' subsistence economy. First, the Missouri River bottomland, with its excellent soil, moisture conditions, and temperature, provided high quality cropland not found elsewhere on the upland plains. Second, the riverine area furnished habitat for game animals as well as pasture and shelter for domestic stock. Finally, the bottomlands supplied the timber necessary for heat, buildings, and shelter. The construction of the Garrison Dam relocated 1,700 people and disrupted the entire social, economic, and cultural life of the tribes involved. The Indians lost 90 percent of their timber and much of their wildlife habitat. The dam flooded ancient burial grounds and religious sites.

When the entire costs were considered, the Army Corps of Engineers paid less than $100 per acre for the Missouri River Indian reservations. As a result of the project, the Indians bore a disproportionate share of the social costs of water development, while having no share in the benefits. Poverty, disease, and shortened lifespans were only a few of the social costs the Indians experienced after their removal to make way for the reservoirs.[11]

Water resource development evaluation methods do not adequately consider cultural costs and values. These values are not limited to ethnic minorities. Rural residents, religious groups, or minority political values may be concentrated in the area used for a major federal water development project. Older people are especially affected by forced relocation. The same principle also holds for scientific values. Many archeological and historic sites have been flooded by water projects. The main stem Missouri River projects wiped out many frontier historic sites, Fort Randall dam flooded 120 historic sites and features, and Garrison Dam buried 77 historic and hundreds of archeological sites. Such resources are not computed in the cost-benefit analysis of water development projects.

Problems Over Time

Congress authorizes water projects with assumed life spans, but it ignores several problems that often emerge over time. First, some projects are outdated before their planned life span is over, due to silting or construction or design errors. Modification or rebuilding can cost more than the initial construction; yet, this is not considered in the cost-benefit analysis. As might be expected, the construction costs of water development projects have been grossly underestimated, while the maintenance and rehabilitation costs have been virtually ignored. All project costs and operational expenses must be clearly planned and considered.

Second, there is the problem of safety. As dams age, their safety decreases. Older construction methods, such as hydraulic earthfill, are known to have a higher risk factor than other construction methods. In the aftermath of the Teton Dam failure, Congress proposed new federal dam inspection programs. With 2.5 million dams and approximately 50,000 over twenty-five feet high, dam safety will probably be the next area of major concern in water development policy.

Third, after several recent dam failures, insurance companies are refusing to insure private dams. Proposed legislation would make the federal government guarantee liability insurance for private water projects. Not only should the insurance costs be part of federal cost-benefit analyses, but it is questionable whether the government should assume the risk of private water development. Such a program not only encourages dam construction through a subsidy, it also discourages safety programs, rehabilitation, and reasonable management behavior.

As a policy issue, federal water development programs have traded a predictable and relatively low risk of flooding for an unpredictable risk enhanced many times over. The damage done by the Teton Dam failure was not just a relatively low level 25- or 100-year flood. It was a massive manmade flood far larger than anything previously experienced. Damages exceeded $1 billion for a project with estimated benefits of only a few million dollars. The same principle is true for the Missouri River. Nonstructural alternatives to lessening flood control were not considered in the Pick-Sloan Plan. Development and construction of the large main stem reservoirs have given residents and officials a sense of security. Such security may be unwarranted. Fort Peck Dam, one of the world's largest, is approximately forty years old, an age when historically there is an increased likelihood of failure. In addition, the construction type (hydraulic earthfill) is now considered the most unsafe method of dam construction (a large section of Fort Peck failed during the initial filling phase in 1938). Garrison Dam immediately downstream has a safer construction method of rolled earth, but it was built on a site that was considered unsafe in early Army Corps of Engineer studies.

Neither the hazards of dam failure nor attempts to consider the costs of failure have been part of federal planning or cost-benefit analysis. The replacement of low level predictable floods by the possibility of massive manmade floods, however remote, is no reason to drop this portion of the equation from consideration.

Accounting of Benefits and Costs

The social costs of water development projects have been the focus of increased discussion and concern. Most recently, the United States Water Resources Council added more detailed and elaborate evaluation of social and noneconomic aspects for water resources planning. This is an important step, but several fundamental problems remain.

First, the methodology to evaluate social accounts and costs has not been agreed upon. This rather basic question comes from not only the lack of experience and methods to evaluate projects, but from a lack of basic data. In addition, new problems such as safety continue to surface.

Second, Congress has authorized many hundreds of water projects that have not been funded. As appropriations become available, or as local Congressmen succeed in getting funding, more projects will be continued or initiated. Congress has decided that the new cost-benefit accounting will not apply to over 800 authorized but unfunded projects. Ignoring new social concerns, Congress funds many projects that were justified using low discount rates or outdated economic analysis. Thus, water development agencies and special interest groups continue to construct economically inefficient projects that promise miraculous results for the community, while experience has shown that these results are more often detrimental in ways that have not been fully considered or explored.

Third, most resource development of water increases the rural to urban flow of resources, economic growth, and political power. The distribution of both benefits and costs over geographic space is an important issue.

Continued Evaluation

One of the most remarkable situations in the benefit-cost process used in federal water projects is the lack of ex post evaluation. There has never been a comprehensive evaluation of benefits and costs, both economic and noneconomic, and their distribution after project completion. No one asks, "Did the project meet its objectives?" Yet, this question is critical. Many of the largest projects were developed with great haste, many were designed as single purpose projects and later modified, and many occupy prime natural resources sites.

Are the nation's water resources being optimized? Without a continual process of evaluation that question cannot be answered.[12] It has been assumed that dams are a fixed resource once they are completed. Yet, changing national, regional, environmental, and social priorities, along with economic shifts in market conditions, may demand the complete redevelopment or dismantlement of a water project. Planning has been static, not dynamic, and water projects that are static cannot optimize the use of public resources. The Pick-Sloan projects may not only be inefficient, they may also represent the least optimal methods of utilizing the Missouri River resources. These projects have a fixed physical life span; yet, their economic life span and efficiency have not been considered in detail. In addition, if the process of water resource development is to be rational, the complex problem of redevelopment and reclamation of land and resources needs to be investigated long before the economic life of a project ends.

Historically, federal water resources planning has focused on predevelopment and development. Yet water projects have a beginning, a middle, and an end. The concept of a continual process of reevaluation is accepted almost everywhere except in federal water resource development.

Conclusion

Environmental problems of federal water projects have been well documented elsewhere; therefore, only a brief summary will be presented here. While water development projects have contributed to water quality and other aspects of the environment, they have also created many problems. Some examples have been the increased upstream and downstream erosion from reservoirs because of shifts in river flow dynamics. Large reservoirs have drastically altered fish and wildlife populations. Some projects have endangered rare plants or animals. Plant and animal communities have been grossly altered in some areas. Large reservoirs have triggered earthquakes and land slides. Dams geologically and biologically alter natural processes and change the environment. The array of biological and physical changes resulting from major projects are seldom fully studied or appreciated prior to construction.

Federal water projects have ignored many of the social and cultural effects of development. Spatial allocation of benefits and costs has not been clearly understood. Most important, the water development process lacks ex post evaluation with continual monitoring of a project's economic life and efficiency. New social, cultural, and political values must be part of the process of development and utilization of water resources. If not continually evaluated, water resource projects become stagnated with outdated facilities and products. Authorized by antiquated standards, federal water projects may become national monuments to archaic water policies.

There are a number of people who benefit from present water development policy. Certainly the immediate influx of construction money from the common pool of public tax dollars neatly accommodates politicians. The benefits are available during the election; the social costs and long-term problems arrive years later. The constituents that benefit are specific, and they have a clear and understandable interest in supporting the existing system. Consultants, engineering firms, and construction companies have a concise understanding of the committee process and appreciate the agencies that propose the projects and the politicians that deliver the appropriations. Bureaucrats have a tenacity and dedication to their goals that pales the most zealous environmentalist.

What will work to change the system? Water development agencies certainly need institutional reform. The shortsighted approach of changing personnel or congressional committee chairmen will probably prove fruitless. One useful reform would be to separate the cost-benefit analysis of projects from the agencies involved. This could eliminate the self-interest of an agency in deciding the size and scope of its own programs. An independent group, such as the Office of Management and Budget, would be better suited to make the initial evaluation.

There should be a more detailed social and cultural accounting of both benefits and costs. Ideally, each cost-benefit evaluation should spell out those who will receive benefits and those who will pay the costs—economically, socially, and culturally. The accounting system should include a visual demonstration of the distribution of both benefits and costs over space or geography. This would help illustrate the equity issues that often arise.

Another important element is the display of social costs and benefits over time. Included would be a regular and complete evaluation to determine the optimal use of water resources. With an updated data base and continual reevaluation, projects would not become outdated monuments to bureaucratic pathology. Redevelopment or the dismantling of projects, large or small, would be planned and anticipated long before the actual demise of the project. With a complete process of evaluation, the opportunity to correct mistakes and write off costs would be readily available.

The social accounting system would be aided by an advocacy resource development approach. Those groups that fear or expect adverse project impacts would be assured the technical data and expertise to examine their concerns and to prepare alternatives. The advocates for minorities, whether cultural, social, religious, or political, would also have the opportunity to present mitigating measures or alternatives to development projects. The advocates would be granted funds that would be included in the project development costs. In this manner, decision makers would have a more complete and comprehensive understanding of water development projects and their social costs and benefits.

Thus, the ideal institutional change would be to separate each project justification from the agency responsible for its construction and management. Implementing an accounting system for benefits and costs that would include all available social and cultural concerns would complement this requirement. An independent advocacy resource development group funded by the development agency and representing minorities that fear adverse project consequences would also improve institutional incentives. The accounting system would consider problems over time and the spatial distribution of benefits and costs. More important, the continual evaluation throughout the life span of every project assures optimal and safe resource use.

With such a system institutionalized into the federal water development process, some of the problems that have been inherent in the process could be reduced. It is unrealistic to expect that the wasteful and inefficient processes of the past will be eliminated. But, if nothing else, more people would be drawn into the "cozy relationship" the water triumvirate has enjoyed.

Notes

1. Harold M. Schmeck, Jr., "Environmental Groups Back Carter Effort to Curb Dam Building," *New York Times*, April 17, 1977, p. 1.

2. U.S., Congress, Senate, Committee on Environment and Public Works, *National Dam Inspection Act Amendments*, 95th Cong., 2d sess., serial no. 95-H43, p. 001.

3. Ibid., p. 42.

4. River Conservation Fund, *Flowing Free* (Washington, D.C.: River Conservation Fund, 1977), p. 21.

5. U.S. Department of Agriculture, Economic Research Service, *A History of Federal Water Resources Programs, 1800-1960,* Misc. publication no. 1233 (Washington, D.C.: Government Printing Office, 1972), p. 3.

6. The best summary of the Pick-Sloan issue is in Marian E. Ridgeway, *The Missouri Basin's Pick-Sloan Plan*, Illinois Studies in the Social Science, vol. 35 (Champaign: University of Illinois Press, 1955).

7. U.S., Congress, Senate, *Appendix Congressional Record*, 78th Cong., 2d sess., 1944, pp. A4381-82.

8. U.S. Department of Agriculture, Economic Research Service, *History of Federal Water Resources Programs*, p. 19.

9. Bernard Shanks, "Missouri River Development Policy and Rural Community Development," *Water Resources Bulletin* 13 (April 1977):255-63.

10. For example, see Richard L. Berkman and W. Kip Viscusi, *Damming the West* (New York: Grossman Publishing, 1973), chap. 7.

11. Bernard D. Shanks, "The American Indian and Missouri River Water Developments," *Water Resources Bulletin* 4 (June 1974):573-79.

12. Two of the few ex post evaluation studies found major errors in both the benefits and costs achieved at large projects. See Robert Haveman, *The Economic Performance of Public Investments* (Baltimore: Johns Hopkins University Press, 1972); and Bernard D. Shanks, "Indicators of Missouri River Project Effects on Local Residents: An Advocacy Resource Development Approach" (Ph.D. diss., Michigan State University, 1974).

A Perspective on BLM Grazing Policy

by Sabine Kremp

The Bureau of Land Management (BLM) administers 447 million acres of public lands in the United States.[1] The largest portion of this land is in Alaska and is only temporarily under BLM jurisdiction. This paper will concentrate on the remaining 171 million acres of public lands in the eleven western states over which the agency has permanent jurisdiction.

One hundred and seventy-one million acres is, to say the least, a large amount of land—more than the combined states of Utah, Wyoming, and Idaho. Most of this land is arid, lies at a low elevation, and has been used predominantly for grazing.

Until 1964, the BLM's major legal objective was to administer grazing privileges on public lands. In 1964, however, Congress broadened the scope of BLM's directive when it passed the Classification and Multiple Use Act. The Federal Land Policy and Management Act of 1976, often referred to as the BLM Organic Act, also provided that the BLM administer public lands under the principles of multiple use and sustained yield. Section 202 reads:

> In the development and revision of the land use plans, the Secretary shall . . . use and observe the principles of multiple use and sustained yield set forth in this and other applicable laws. . . .

It is apparent that the BLM intends to meet its management objectives through widespread implementation of specialized grazing systems, especially rest rotation. The BLM revealed its preference for specialized grazing systems when it released an environmental impact statement (EIS) on its national grazing program on December 31, 1974.[2] The agency reaffirmed its choice in its 1975 Range Condition Report for the Senate Committee on Appropriations.[3]

The adequacy of the 1974 BLM National Grazing EIS was challenged in court by the National Resource Defense Council and other conservation groups. As a result, the BLM is in the process of preparing 153 EISs on specific planning units in the eleven western states.[4] About half a dozen EISs were released by April 1978; none has yet been accepted by the Council on Environmental Quality.

Four BLM grazing EISs were reviewed for this paper: the Challis EIS, the Uncompahgre EIS, the Rio Puerco EIS, and the San Luis EIS.[5] Each statement recommended instituting specialized grazing systems on 80 to 90 percent of the public land involved.

After reading these EISs, one is left with the impression that the public can have more red meat, more wildlife, more recreational experiences, a healthier range, and an improved watershed by simply funding the BLM. All the agency needs to do, we read, is construct enough fences to convert the open range into pastures, develop a water source for each pasture (usually by burying pipelines), and require that livestock be regularly shuffled back and forth between the pastures. Far from being a panacea, however, the evidence suggests that a blanket policy of specialized grazing management may not only be economically unsound, it may also contribute to environmental degradation.

This paper will trace the development of the proposed grazing policy, summarize some of the major literature on specialized grazing systems, and raise questions about the environmental, economic, and social desirability of implementing the BLM's proposed grazing policy.

A Short History of the BLM

After the Revolutionary War, the new nation rapidly expanded its western frontier, beginning in 1803 with the Louisiana Purchase and ending with the Alaska Purchase of 1867. The Gadsden Purchase in 1853 annexed parts of Arizona and New Mexico and marked the last boundary change of the contiguous United States.

As it was bought or won, the dominant political policy assumed that these newly acquired federal lands would be sold or granted to states, businesses, or individuals. The disposal of these lands would strengthen the federal treasury, aid in the settlement and defense of the western boundaries, and provide the citizens of the new nation with the means for securing their own economic livelihood. The Congress of the early and middle nineteenth century did not envision a nation with one-third of its land in public ownership.

In April 1812, Congress established the General Land Office as a part of the Treasury Department "to survey, manage and dispose of public lands."[6] The General Land Office administered various ordinances that granted land to settlers until March 1849. At that time, the General Land Office, Office of the Census, and Office of Indian Affairs were brought together as the Department of the Interior. The new department continued to administer land grants and land sales throughout the last half of the 1800s and into the first half of the 1900s.

Around the turn of the century, however, a new philosophy regarding the disposal of public lands began to emerge. Much of the forested lands in the East and Midwest, especially land that had been acquired cheaply or that was still under public ownership, had been stripped of its timber by private exploiters and left to erode and wash away into the nearest river.[7] Political support mounted for the reservation and protection of some public land and, in 1872, the first large piece of land that was to be retained by the public into perpetuity, Yellowstone National Park, was established. Shortly thereafter, in 1891, Congress passed the Forest Reserve Act, authorizing the president to withdraw public land for the protection of trees. By 1901, 148 million acres had been set aside for this purpose, largely through the auspices of Teddy Roosevelt and Gifford Pinchot. In 1905, the Forest Service was established to administer the land.

Throughout the early 1900s, millions of acres were set aside for national forests, national parks, national wildlife refuges, national monuments, and national historical sites. Additional millions continued to be granted or sold to settlers, states, and businesses.

During this time of intense disposing and reserving, the government ignored a large portion of land, the hot, dry, dusty, rocky, "desert" land so predominant in the West. This has been labeled by historians as "the land nobody wanted." Settlers did not claim this land under the various Homestead Acts because its productivity was simply too low to make the cost and the effort of homesteading worthwhile. In addition, the annual taxes that would have to be paid discouraged its transfer into private ownership. The federal government did not want the land for any particular purpose either. Because it was mostly unforested and not considered particularly spectacular or useful, the land was not included in the national parks or forests.

The public land did, however, have value as marginal grazing land. Unclaimed by anyone, it became essentially a huge common property pasture. Cattlemen and sheepmen started grazing their livestock on the free, unregulated forage by the 1850s.[8] Because no property rights had been defined and because no management scheme had been set up to regulate the use of the range, each livestock operator had every incentive to contribute to the overutilization of the public range land. The benefit of adding another animal to this land accrued to the individual operator, while all range land users shared the costs that the additional animal imposed. As long as the benefits to the stockman of adding one more cow or sheep outweighed his share of the costs, the rational livestock operator would continue to add animals to his herd. Garrett Hardin succinctly states the common property problem:

> Therein is the tragedy. Each man is locked into a system that compels him to increase his herd without limit—in a world that is limited. Ruin is the destination toward which all men rush, each pursuing his own best interest in a society that believes in the freedom of the commons.[9]

It is important to note that an awareness of the situation would probably not have stopped the unregulated western public range land from being damaged by livestock operators. For each individual who refrained from adding more animals in the interest of the common good, another operator would have added that many more. With the absence of a universal altruism, the feeling of "if I don't use it somebody else will" prevails in a commons situation.

By the turn of the century, most of the public range was extensively deteriorated and eroded because of heavy overgrazing. Only those areas that were poorly watered or otherwise inaccessible to livestock use remained in good condition.[10]

Range conditions continued to decline on the public domain lands in the early 1900s. Finally, in the 1930s, ranchers and others recognized that something needed to be done before the West became a vast, useless dust bowl. Transfer of the grazing lands to private ownership would have alleviated the problem to a certain extent by providing incentives for determining the optimal stocking rate to each ranch owner. The benefits of adding another animal would still go to the ranch operator, but the costs of an additional animal would (in the absence of external costs) no longer be shared. If a rancher over-grazed his land, he and his heirs would personally suffer from the reduced value and productivity of his property. So, with private ownership, there would have been fairly strong incentives to encourage the maintenance of the productive capacity of the land.

However, individual ownership had not been established for this land through the various Homestead Acts. The immediate alternative, then, was to set up some type of governmental agency that had the power to regulate the use of the public range. In June 1934, Congress passed the Taylor Grazing Act, which established the Grazing Service in the Department of the Interior. The agency's mission was to administer the range land and stabilize the western segment of the livestock industry.

From the start, the agency was underfunded. Secretary of Interior Ickes estimated in 1939 that it would cost $150,000 annually to manage 179 million acres.[11] Even in an era of uninflated dollars, this was an unreasonably low amount of money. With a low budget and an almost exclusively western clientele, Grazing Service administrators often found it difficult to lower livestock numbers to within the carrying capacity of the land.[12] In addition, the agency did not have the political clout to charge stockmen full value for grazing privileges.[13] By 1946, the agency was being condemned on all sides. Influential western senators were angry that the Grazing Service was attempting to raise grazing fees to market value. Eastern representatives were angry because the agency failed to raise grazing fees to market value. That year, Congress consolidated the Grazing Service and the General Land Office into the Bureau of Land Management (BLM). The fledgling agency was granted only half of the previous budget, and appropriations for fiscal 1948 were even lower than for 1947.[14]

Initially, the BLM was handicapped not only financially, but also administratively. Congress failed to grant an organic charter to the agency, and it was required to operate under the guidance of some 3,000 often contradictory laws.[15] Finally, on October 21, 1976, Congress passed the Federal Land Policy and Management Act of 1976, often referred to as the BLM Organic Act. It authorized, among other things, that the public lands be retained under federal ownership, thus making the BLM a permanent resource management agency. It also provided a new organization chart for the responsibilities of the BLM, and it mandated that the agency continue to administer the land under the principles of multiple use and sustained yield.

The Condition of the Public Lands
and the Importance of Grazing

In 1968, Pacific Consultants completed an analysis of range conditions on public lands for the Public Land Law Review Commission. They found that 19 percent of BLM land was in good or excellent condition, 52 percent was in fair condition, and 29 percent was in poor condition.[16] Another study was conducted in 1974, when the Senate Committee on Appropriations asked the BLM to prepare a nationwide assessment of range conditions. This report indicated that about 17 percent of range lands were in good or excellent condition, 50 percent were in fair condition, and 33 percent were in poor or bad condition. The reported trend was 19 percent improving, 68 percent static, and 16 percent declining.[17]

Although these studies indicate that the public domain range has improved since 1936, when only 1.5 percent was considered to be in good or excellent condition and the rest was classified as unsatisfactory,[18] BLM administered range land is not considered to be in as good condition as adjoining private land or national forest lands.[19]

The BLM states that its own ranges are not being properly managed and that conditions are declining and will continue to decline under the present system.[20] In 1974, the BLM state office in Nevada printed a critical in-house self-evaluation entitled, "Effects of Livestock Grazing on Wildlife, Watershed, Recreation and Other Resource Values in Nevada." The report showed that the BLM was understaffed in the state and that the range lands were severely undermanaged.[21] The Nevada study was followed by several other state studies, all of which pointed out a lack of personnel and indicated that the proportion of money being spent on range management was declining. For example, the Colorado study showed that, after adjusting for inflation, only about 50 percent as much money was spent on range management in 1974 as in 1965.[22]

In February 1976, Thomas Kleppe, director of the BLM, told the Society for Range Management that 83 percent of public lands were in "something less than satisfactory condition."[23] The BLM has pledged that "first and foremost, range policy will be geared to improving all resource conditions to acceptable levels."[24]

About 273 million acres of federal lands are currently grazed, of which the BLM manages approximately 150 million acres. About 25,000 operators hold grazing permits for use of federal lands,[25] and approximately 20,400 of these stockmen graze their animals during part of the year on BLM administered lands.[26]

The federal range land (including Forest Service lands and other federal lands) provides about 3 percent of all forage consumed by livestock in the United States and about 12 percent of all forage consumed by livestock in the western United States.[27] Although these figures indicate that forage from public range is not too important to the livestock industry, they are somewhat misleading. According to the Public Land Law Review Commission, the public lands are

> often crucial to individual ranch operations, supplementing the feed of private lands by supplying seasonal grazing. Without the privilege of grazing public lands, many ranches would cease to exist as economic units, or would be forced out of business due to the high cost of substituting other sources of feed. The western range livestock industry, which is built around the public lands, also must be viewed as an important source of range livestock for feeder lots throughout the West and Midwest.[28]

Even though a segment of the livestock interest apparently believes that "most of this natural range land is good for nothing else but producing beef,"[29] the dominance of grazing use on BLM lands is being increasingly challenged by other natural resource users, especially recreation, mining, and wildlife interests. These groups often perceive grazing policy as conflicting with their own demands on public lands. In attempting to improve all resource conditions to acceptable levels, the BLM is under pressure to repair deteriorated range conditions, provide constant or increased quantities of forage to livestock permit holders, and accommodate the often competing public demands for the use of these federal lands.

It appears that the way the BLM intends to improve the public lands, meet its management objectives, and satisfy the various public uses is through widespread implementation of specialized grazing systems.

Specialized Grazing Systems

Very generally, specialized grazing systems are categorized in this paper as either rest rotation or deferred rotation grazing. Under either category, the range is fenced into pastures of approximately equal carrying capacity, and each pasture is provided with a water source. According to Robert E. Steger, rotation grazing implies the progressive use of several pastures so that no pasture is used during the same season for two years in a row.[30] Each pasture is grazed and then not used. Over a given period of time, all pastures receive a grazed period followed by a period of nonuse. Under rest rotation, one pasture in a range is not used for a full year. Under deferred rotation, one of the pastures is not used during the growing season or for part of a year. Under either type of rotation grazing, the stocking rate on the pastures used will usually be heavier than under continuous grazing. (Continuous grazing refers to the system where cattle or sheep are grazed on the same land year after year.)

Although rotation grazing is generally considered a relatively new management technique, the concept originated in the nineteenth century. Jared Smith wrote in 1895:

> Clearly then, if the grazing quality of the ranges is to be improved, they must be so treated that the nutritious native species of grasses and forage plants can spread by means of ripened seed. This can be accomplished by dividing the range up into separate pastures and grazing the different fields in rotation.[31]

And in 1899, he said:

> Less than thirty years ago, 4,000,000 buffalo and countless numbers of wild horses roamed unrestricted over the region in question, gradually moving northward as the season advanced, returning southward at the approach of winter. This natural movement of the stock permitted alternation of pasturing and rest for the land, resulting in the maintenance of the forage supply.[32]

Nevertheless, August (Gus) Hormay, not Jared Smith, has been christened the "father" of rest rotation management. Hormay was a range management advisor to the BLM and initiated research in testing rotation theory.

Hormay explains the rationale for implementing rest rotation schemes in a small pamphlet entitled *Principles of Rest Rotation Grazing and Multiple-Use Land Management:*

> Assumptions that plants can be grazed to a proper level through regulation of stocking is unrealistic because of the grazing habits

of livestock. Livestock graze the range selectively, by species and areas. They consistently graze the more palatable plants and accessible areas closely, and invariably, beyond proper-use levels. The pattern of use is very uneven, but much the same from year to year. . . . So under continuous grazing at any stocking level, the more palatable and accessible plants are gradually killed out. Livestock then graze on less desirable plants. This process leads progressively to ever enlarging areas of deterioration. . . . The better forage plants and all others can be maintained, however, by periodically resting the range from use.[33]

Not all scientists are as enthusiastic about rest rotation as Hormay, however. A team of scientists reviewed the draft EIS for the Challis Planning Unit and concluded that

rest rotation grazing has seemingly been adopted without thorough consideration of the advantages and disadvantages. . . . Rest rotation is unfortunately not a panacea to all grazing problems. In the long run, use of this system may be highly detrimental to both the range livestock industry and the condition of the range.[34]

In rest rotation management, a grazing allotment is divided into two to five or more pastures. The total number of pastures should depend on the size and condition of the allotment. A three or four pasture system is typical. In one grazing season of a three section allotment, the first section is intensively grazed the entire period, the second section is grazed after forage grasses have gone to seed (so that seeds will be trampled into the soil), and the third section is rested for the entire season, allowing new forage to develop. The following year the second section is grazed, the first is rested, and the third is used only after grasses have gone to seed. Stocking rates on grazed pastures are based on utilization of all forage species. Fairly heavy stocking forces greater use of the less palatable forage species and the less accessible grazing areas, thus giving no plant species a competitive advantage.[35] The same number of cattle often graze an allotment under rest rotation grazing as under continuous grazing systems. Thus, a pasture in a rest rotation scheme might be stocked two or three times as densely as under continuous grazing.

Specialized Grazing and Multiple Use

The BLM is mandated to manage the public lands under the principles of multiple use and sustained yield. The Federal Land Policy and Management Act of 1976 (Section 103) defines multiple use as

> the management of the public lands and their various resource values so that they are utilized in the combination that will best meet the present and future needs of the American people; ... with consideration being given to the relative values of the resources and not necessarily to the combination of uses that will give the greatest economic return or the greatest unit output.

The Act defines sustained yield as

> the achievement and maintenance in perpetuity of a high-level annual or regular periodic output of the various renewable resources of the public lands consistent with multiple use.

BLM grazing policy must, then, be consistent with these two principles.

A major selling point that the BLM uses in trying to get public support for its grazing schemes is that its proposed systems will benefit most major users of the public range. According to the recently released Uncompahgre Basin Resource Area Grazing Environmental Statement, where 84.3 percent of the land is to be managed with specialized grazing, the environmental impacts of the proposed plan would be

> runoff and erosion would decrease as a result of improved watershed conditions; vegetation would improve in quantity and quality; terrestrial wildlife habitat would improve, resulting in population increases; aquatic and riparian habitat would improve or stabilize; ... livestock forage would increase; area income would increase. Overall visual quality would not be reduced by range improvements. Impacts on recreation would be negligible.

> Short term adverse impacts would include reduced vegetative vigor and reproduction due to spring and summer grazing and to range improvements; and increased wildlife-livestock competition on critical winter ranges. Long-term adverse impacts would include some stream bank deterioration by grazing and tram-

pling; and some cultural sites damaged by cattle rubbing and open to vandalism.[36]

Let us examine some of the major categories of multiple use that the BLM claims will benefit from implementation of specialized grazing systems on public lands.

Effects on Desirable Forage Production

Many studies on vegetational responses to rest rotation or deferred rotation grazing indicate that plant species that are considered desirable for domestic livestock increase in quantity and vigor.[37] A few studies, however, indicate that desirable vegetation decreases or is unaffected by rotational grazing schemes.[38]

Most of the grazing studies that either support or refute the advantages of rotation management can be criticized for one reason or the other.[39] Some major criticisms are:

1. Long-term effects of specialized grazing systems can be determined only by running long-term experiments. Due to the expense and other constraints, most grazing studies have been conducted for relatively short time periods.

2. Adequate controls are needed if any conclusions are to be drawn concerning the effect of specialized grazing systems. These controls are lacking in many of the studies.

3. There has been a tendency to credit all vegetational improvements to rotation grazing schemes alone, although brush control and reseeding might have been implemented in the allotment.

4. Range conditions vary extensively from area to area. A system that will work at one location will not necessarily work at other locations. Yet, results are often generalized.

Most range scientists agree that rest rotation or deferred rotation systems can improve range quality under certain circumstances. However, as indicated above, there is no guarantee that a blanket policy of intensive grazing management is likely to work.

A single rest rotation grazing scheme, for example, while recognized as an excellent management tool in some situations, cannot possibly be suited for all ecological types—e.g., for desert as well as mountain range. It must be considered for use where appropriate along with many other possible improvement practices. We are particularly concerned at the heavy reliance that is currently being made on "recipe type" grazing systems. For instance, the allotment management plans on all the BLM pilot areas being evaluated as part of the Natural Resource Defense Council suit are equating a Hormay type rest rotation scheme with good, intensive management. This type of system simply has not been adequately tested in most vegetation types and has a great potential for range destruction as well as range improvement.[40]

The effects of rest rotation grazing have been most thoroughly studied in the Harvey Valley of California. This part of California is wetter and higher than the desert shrub and grassland range of much of Utah, Nevada, Arizona, New Mexico, and southern California. Annual precipitation in the Harvey Valley is eighteen inches, and elevations vary from 5,600 feet to 7,500 feet.[41] Although rest rotation management was successful in the bunchgrass range of Harvey Valley, it cannot be assumed that this management technique will benefit the range in other ecosystems. And yet, rest rotation and deferred rotation management are currently planned for much of the desert range land.

It takes about seven years for the vegetation of a desert shrub range to recover after being subjected to a season of intensive grazing.[42] Where the BLM has already implemented rest rotation management on the desert ranges, it has not allowed for seven years of rest.[43] The effects of this scheme on the desert still remain to be determined.

The BLM claims that vegetation will improve with specialized grazing systems not only because of the opportunity to grow, but also because livestock will help scatter seeds. Again, this claim has not been substantiated by research.

Effects on Domestic Livestock

There is a great deal of disparity in the literature concerning the effects of rotation grazing on livestock. For example, Driscoll reviewed twenty-nine studies reporting weight gains or losses of cattle as a response to specialized grazing systems.[44] The results were summarized by Shiflet and Heady.[45] Eight studies reported greater gains for specialized systems, twelve reported greater gains for continuous grazing, and nine showed no appreciable differences. Since no two studies were alike in terms of area or design, no statistical comparisons could be made.

The response of livestock to rotation grazing schemes is, of course, very important to the success of instituting this range policy and of obtaining ranchers' backing. These management systems are somewhat complex and require ranchers to spend more time herding livestock and maintaining fences. The agency apparently hoped that the promise of improving the range without significantly reducing stocking levels would appeal to the livestock operators holding grazing permits for public land. However, some of these operators are highly critical of the proposed changes. They say that specialized systems will cost them more without bringing better results. They contend that maintaining new fences and water developments will be expensive and that herding livestock from pasture to pasture will take more time and money while simultaneously causing the animals to lose weight from the increased movement.[46] Other ranchers and agency personnel maintain that within a few years, once range conditions have improved and the animals have gotten used to the idea of moving every now and then, livestock will show greater weight gains in comparison to animals grazed continuously.

The latter point of view is supported by several studies on rest rotation management.[47] Economic studies also indicate that ranchers will benefit economically from BLM implementation of intensive grazing schemes.[48] These studies, however, assume that the BLM will pay for the capital costs of fence construction and water development that are necessary to implement the grazing systems.

Watershed Protection

In most cases, the BLM asserts that watershed quality will increase and soil erosion will decrease with rest rotation and deferred rotation management. The agency bases its claims on the assumption that its management scheme will increase the quantity of forbs and grasses on the range. Although the BLM acknowledges that cattle crowded into a pasture will cause soil compaction and thereby decrease the rate of water infiltration, this decrease would be more than offset by increased infiltration resulting from a more extensive vegetative cover. However, according to Gifford and Hawkins:

> Published evidence fails to show that any grazing system consistently or significantly increases plant and litter cover on watersheds. . . . For all practical purposes there is no information

available on the impact of rest rotation systems on watershed cover, even though rest rotation systems are being advocated for much western rangeland.[49]

Several of the EISs examined claim that the proposed grazing systems would decrease erosion and peak surface water runoff even in the event of high-intensity summer thunderstorms. The San Luis Resource Area EIS, however, asserts that watershed conditions would improve overall, but that "in the event of infrequent, high-intensity storms, sediment yield would increase by 3 to 5 percent on allotments receiving periodic, concentrated grazing during spring and summer."[50]

The combined effects of rest rotation or deferred rotation grazing and high intensity storms are only beginning to be researched.[51] Under the BLM grazing schemes, livestock are forced to eat all available forage. In addition, heavy stocking rates cause soil compaction, which reduces the infiltration rate of water. It seems highly possible that sheet erosion could occur if a high intensity storm hit a pasture that has not had time to recover from intensive grazing.[52] Gifford, Hawkins, and Williams estimate that land must be completely rested for four years after specialized grazing to fully recover its original infiltration rate. Most of the systems proposed by BLM provide for only two years of rest.

Wildlife

There is much speculation in the grazing EISs concerning the effects of rest rotation and deferred rotation management on wildlife habitats and populations. A literature search reveals, however, that there is very little hard data to support the various contentions.

The BLM has reserved a certain number of AUMs for terrestrial wildlife in each planning unit. (An AUM is an animal unit month, defined as the amount of forage required to sustain the equivalent of one cow, one elk, five sheep, or five deer for one month.) The agency assumes that, although specialized grazing schemes will have some detrimental effect on a few individual animals, wildlife populations in general will benefit because of the increased forage and developed water sites.

Some wildlife specialists, however, argue that the well-being of many species will almost certainly decline. Fencing restricts the movement of big game animals. In addition, if rest rotation works, shrubs are largely replaced with grasses. Deer, sage grouse, and other species dependent on shrubs may be severely affected by the elimination of sage and bitterbrush. The negative sentiment of some wildlife biologists is expressed in a personal letter dated January 3, 1977, from Glen Griffith, director of the Nevada Fish and Game to director of the Nevada BLM:

> there is no alternative for the Nevada Fish and Game Department except to heartily condemn the (grazing) proposal as being detrimental to wildlife and not meeting the test of the "Organic Act" definition of land management for multiple uses.

The BLM cites research in its Uncompahgre EIS, indicating that

> deer are creatures of habit and will return to the same areas year after year. Because of this tendency, deer have been known to starve to death in an overutilized area almost within view of adequate forage. This is not expected to occur in the Environmental Study area because of the rest that will allow browse to recover from livestock grazing.[53]

The EIS does not tell us, however, where the deer will go during the time a pasture is maintaining or recovering from two or three times the normal number of cattle or sheep.

Nor is it at all certain how elk will fare under specialized grazing systems.

> The preliminary findings of the Montana Cooperative Elk-Logging Study indicate that elk will not go into an area where cattle have been and will move out of an area as cattle move in. If this study fully applies to the Environmental Study area, the majority of the elk use would be in the rested pastures with some use of the other pastures because of the tendency of elk to follow the same routes to the same areas year after year.[54]

This leads to a very important point not addressed by the agency. If deer and elk act like they are "supposed" to and utilize forage in the pastures being rested, what happens to the effectiveness of the grazing systems? On some planning units, a rather large proportion of AUMs is being reserved for wildlife use (30 percent in the Challis unit and 35 percent in the Uncompahgre Unit). If these pastures are heavily grazed by wildlife after livestock move out, then they are certainly not being rested.

Specialized grazing systems require that the open range be crisscrossed with fences. According to the Challis EIS:

> While construction standards would be followed to permit passage of most big game animals, a few animals would continue to get caught in the fences. This would cause some additional mortalities of antelope and deer, and perhaps other wildlife, that would not have occurred without the fences.[55]

Fences will have an impact on wildlife populations, but studies have not been done on the magnitude of the impact.

Fisheries biologists are also concerned about the possible effects of rest rotation and deferred rotation grazing. The well-being of fish and other cold water stream organisms decreases as the sediment load of the stream increases.[56] There is some concern that including fragile streamside habitats in intensive grazing systems will damage the water quality of the fisheries. The Uncompahgre EIS states:

> Site specific grazing systems designed to improve the condition of range plants are not tailored to the physiological requirements of woody riparian plants. The success of rest rotation systems in improving range vegetation does not guarantee that riparian plants bordering a stream within pastures would be maintained.[57]

Riparian vegetation benefits fish and aquatic insects by providing hiding places, by reducing stream erosion, by slowing current velocities, by mechanically holding soil particles, and by maintaining moist conditions in the soil. Consequently, a loss of riparian vegetation and increased trampling of streamside banks could have severe impacts on the quality of the fisheries.

In spite of the controversies and the lack of good data, the BLM assures us in its grazing EISs that wildlife populations will benefit from implementation of its proposed grazing management system. This conclusion seems open to debate.

Recreation

The BLM generally assumes that rest rotation and deferred rotation management schemes will have a negligible or a beneficial effect on recreational experiences. However, as is typical of most of the issues discussed so far, there are insufficient data on which to base any firm conclusions.

For example, sportsmen are supposed to benefit because of larger wildlife populations that will allow for more hunters and increased hunter success. Even ignoring the controversy of whether game populations will actually benefit, increased hunter success does not automatically insure a "better" recreational experience.[58] It is not unlikely that the increased irritation at having to figure out how to cross a labyrinth of fences would more than offset the pleasure resulting from an increased chance of shooting a deer.

Hikers, four-wheel drive and motorcycle enthusiasts, and snowmobilers are likely to be disadvantaged by specialized grazing schemes because the required fencing is an obstacle to cross-country travel, is unsightly, and changes the wild, open range atmosphere. In addition, campers may find themselves competing for campsites with droves of cattle who are packed together more densely than under open range conditions.

Consequently, there does not seem to be any basis for assuming that rest rotation and deferred rotation will have negligible or positive impacts on recreation. Quite to the contrary, recreational experiences are likely to suffer from these schemes.

Summary

The BLM is legally required to manage the public lands for multiple use, and the agency claims that its proposed grazing management schemes are consistent with these principles. As indicated, however, there is little scientific foundation on which to base such a claim. It seems just as reasonable to assume that major resource uses of the public lands will be adversely affected.

Even if the objectives of multiple use are disregarded for the moment, other factors should be taken into consideration when analyzing the BLM grazing policy.

An Economic Analysis of Specialized Grazing Schemes

Most of the literature dealing with rotation management is concerned with the positive or negative biological consequences of the system. The economic aspects have been relatively neglected, and yet fairly large capital investments are required to initiate this type of management. Thousands of miles of fence must be constructed across the open range, often at a cost of $3,000 to $4,000 or more per mile. In addition, each pasture must be provided with some type of water source. The BLM estimates that the initial cost of implementing intensive range management programs on public lands will be about $328.6 million. In addition, annual maintenance costs are calculated to be an extra $33.3 million.[59]

Costs, of course, tell us nothing until they are compared to benefits. If the benefits of the proposed grazing management schemes would outweigh the costs, and if the distribution of these benefits and costs were acceptable to the general public, then all would be well. In spite of BLM's assurance that benefits are greater than costs, there is strong evidence indicating otherwise. The distribution of costs and benefits will be discussed in more detail later, but it is fairly obvious that the beneficiaries of specialized grazing systems are BLM officials and, in some cases, livestock operators. The costs are to be met to a certain extent by livestock permittees, but will undoubtedly be primarily spread over the tax-paying public.

Rozell, Ching, and Hancock analyzed fourteen allotments managed with rest rotation systems in Nevada.[60] They projected an expected increase in forage production and grazing fees over the fifteen-year planning period of the allotments. They compared the costs to the BLM for implementing the grazing systems with the returns that the agency received in terms of increased grazing fees for additional AUMs produced. They then computed the internal rate of return, which is the discount rate that would make the present value of the income stream of a rest rotation system equal to the present value of its cost. According to the authors,

> The internal rate of return was computed for 11 of the 14 observations (allotments). Three observations had to be deleted because net returns for each year over the entire planning horizon were negative. The magnitude of the internal rate of return on the remaining 11 observations ranged from −0.18 to 0.42. The mean was 0.01.

This means that if the BLM were acting solely as a proprietory agent in the interest of the federal treasury, it would have made economic sense for them to implement rest rotation management on these eleven allotments only if they could borrow money at an interest rate of 1 percent or less.

Workman and Nazir calculated similar results in analyzing

twenty-four BLM rotation grazing systems throughout the western states.[61] They found that for these areas, the internal rate of return (considering dollar costs and benefits to the agency only) was 2.37 percent.

Management decisions regarding these public lands should not be made solely by the BLM as a proprietory agent. The internal rates of return on public investment that use grazing fees as the estimate of benefits may be useful for indicating financial feasibility of rotation grazing systems from the viewpoint of the governmental agency. However, since the grazing fee underestimates the full value of the additional grazing by a sizable amount (50 percent or more),[62] using such fees as benefit estimates seriously understates the internal rate of return that accrues to society, the rancher, and the government. When making its management decisions, the BLM assumes that it is responsible to society at large and thus attempts to account for all perceived private and public benefits and costs. In addition, the BLM is mandated to take into account the effect its actions have on the stability of the livestock industry and on national, regional, and local well-being.

In an instructional memorandum, each BLM office is advised to give the "best possible dollar estimate of per-unit benefits" of non-market values, and "feasible, available professional research and other sources of information should be studied in selecting estimates of nonmarket benefit levels."[63] This memo also instructs the agency to include increased rancher profits due to rest rotation as a benefit. Costs to be considered are agency expenses, nonmarket costs (such as decreased wildlife or increased soil erosion), and increased costs to the ranchers.

Since the agency asserts that almost all nonmarket values of the public lands will benefit from intensive grazing systems and that ranchers will, in general, increase their profits through the institution of the proposed systems, it is hardly surprising that the BLM arrives at positive benefit-cost ratios in its grazing EISs. This is true in spite of the high capital investment costs of specialized systems.

However, we have seen that there is no solid basis for these optimistic assumptions. It appears, in fact, reasonable to assume that implementation of a blanket policy of rest and deferred rotation grazing systems may result in the destruction of nonmarket values of the environment. In that case, the American taxpayer will find himself in the position of paying for the deterioration of the western range.

Stabilization of the Livestock Industry

One BLM objective is to stabilize the livestock industry in the West. One approach used is granting subsidies. When ranchers pay less for the forage generated from the grazing program than its net cost to society, then the grazing is being subsidized.

The livestock industry only pays a small part of the costs of implementing the proposed grazing systems. The EISs do not specify who will pay the rest; they simply assume that the necessary funds and personnel will be made available.[64] Since the BLM cannot pay for the capital improvements and extra personnel necessary to implement specialized grazing from the income generated by grazing fees, one can only suppose that a large percentage of the money will come from the common pool treasury.

There is nothing inherently wrong with paying subsidies. Throughout our history, we have become an increasingly redistributive-oriented society. However, when subsidies are paid, they should be acknowledged as such. Nowhere in any of the EISs reviewed is this subsidy explicitly mentioned. It may well be that the American public wishes to keep private western ranches in operation. But without knowing about the subsidy, they cannot exercise their choice through the political arena.

Stabilization of the livestock industry does not necessarily imply that subsidies must be paid. It may well be accomplished by giving permittees greater control and more flexibility over range use, without subsidies. According to the Public Land Law Review Commission, "If more precise standards of permitted use for the maintenance of range conditions are incorporated into permits, the objectives of more certainty in tenure and greater permittee control over range can be obtained."[65]

In essence, the livestock operator could acquire property rights to the land. Constraints would still be imposed on range improvement technology and on the number of livestock that could be grazed. Grazing rights of ranchers who stayed within their constraints would be renewed indefinitely, while the rights of those who violated the constraints would be revoked. Ranchers staying within the stated constraints should receive full compensation for their loss if grazing rights are terminated.[66] The Federal Land Policy and Management Act of 1976 provides for increased tenure of grazing permits and for compensation for their loss. This is probably a step in the right direction.

In the absence of creating more secure property rights, some improvements in the allocation of grazing permits can be accomplished by eliminating eligibility requirements and raising grazing fees. To be eligible to receive a grazing permit, a rancher must have acreage on commensurate base property in the area of reasonable proximity to the public lands. Base property will qualify as commensurate only if it depends on BLM grazing land to round out a year-long livestock operation. In addition, ranches that used the public lands for grazing before the Taylor Grazing Act was passed were eligible to receive permits under a "prior-use" privilege doctrine. These requirements served as a rationing device to allocate permits, because fees were set below the equilibrium price of the forage. Since many livestock operators could not qualify for grazing under these criteria, there was a misallocation of grazing permits. Permits did not necessarily go to those operators who could make the most efficient use of them.[67] If permits were now auctioned to the highest bidder or grazing fees were raised to market clearing levels, the eligibility requirements could be eliminated and the misallocation avoided.[68]

If the grazing fee were raised to the value of the forage of the most efficient users, only they would be interested in utilizing the permits, and greater economic efficiency would result. It must be pointed out, however, that the raising of fees would impose significant losses on current permit holders. A combination of pricing below the value of forage to the stockman and of granting permits of ten-year duration (which could be renewed and transferred) meant that the difference between the fee and the value of the forage was capitalized into permit values. The permits represented a wealth windfall to the original permittee. Successive permit holders, however, had to pay the transfer price for the grazing privilege in order to acquire it. It follows that, if fees were raised to the value of the grazing to the most efficient users, the value of the permit to the ranchers would approach zero, since there would be no differential on which to capitalize. There would then be a transfer of wealth from the permit holder to the government.[69]

Conclusion

Grazing is generally not considered a harmful use of the public lands. Range plants and herbivores have coevolved so that plants are not harmed by some grazing. In fact, ranges can be improved under proper grazing management.[70] Overgrazing, however, is extremely harmful to range land.

Most range scientists agree that specialized grazing systems can improve the quality of some range land. However, they also agree that flexibility in management is essential for good results. Shiftlet and Heady write:

> Specialized grazing systems have a place in range management, but they are not a panacea that will solve all the problems now confronting range managers and operators. A grazing system that depends on manipulating seasonal use should not be confused with a whole management program. A review of the literature plus considerable field experience leads us to view specialized grazing systems with a certain amount of skepticism and to question, in particular, their wholesale use. A highly successful design in one area or even in one ranch is no guarantee of similar success in another location having different vegetation, climate, growing seasons, and managerial objectives. Only after close study and adjustments to meet local conditions can any grazing system be recommended for a specific region.[71]

A range that is already in good or excellent condition does not need rotation grazing and will probably deteriorate if these systems are instituted.

> After a range has been improved to a desired condition, the system may need to be changed or readjusted. Simply stated, adjustments will be required as the key forage-producing species changes.[72]

The BLM does not indicate in its grazing EISs whether specialized grazing schemes will end when (if) the range has improved to the desired level. The agency also does not indicate why it will apply rotation schemes to pastures that are already in good condition. The BLM appears determined to institute specialized grazing systems on as much land as possible for an indefinite time period.

Why is the BLM proposing range management schemes that are likely to be environmentally and economically costly? Are they, as one critic of the Challis EIS suggests, trying to put a "strangling noose upon grazing and furnish ammunition for a slaughter from environmental groups that oppose grazing on public lands"?[73]

The BLM has adopted these schemes on a wholesale basis for several reasons. The agency is under fire to improve the condition of the public range, to provide for a wide variety of often conflicting uses of the public lands, and to manage for multiple use. Under these pressures, rotational grazing schemes easily take on the nature of a panacea. In addition, the uniform application of rotation grazing systems reduces transaction and information costs to the government and makes these lands relatively easy to administer.

Another reason is that the agency itself would benefit. Specialized grazing systems are very expensive, and their implementation would require the BLM to expand its staff and make costly capital investments for fencing and water developments. An expanding agency gives its officials increased discretionary control over resources. A bureaucrat usually has a fair amount of freedom in deciding how new funds will be spent. Work place amenities, such as office furniture and carpets, tend to improve as money increases. In addition, an expanding budget usually provides new jobs, and a bureaucrat's power increases when he can hire new people. Since under civil service laws it is almost impossible to fire anyone, the easiest way for an official to gain control over subordinates is to offer rewards of promotions or improvements in the size of their staff.[74]

From this perspective, it is easy to understand why the Bureau of Land Management is supporting rotation management schemes with such enthusiasm. The agency predicts that there will be benefits to the range, to society, and to themselves.

Suggestions

It is not within the scope of this paper to recommend alternative grazing systems. However, a few guidelines to developing a general approach to range management can be offered.

A blanket policy of instituting grazing schemes is likely to do more harm than good. Grazing systems should be flexibly applied and individually adapted to each area. In many cases, management techniques that are less expensive than rotational systems would achieve grazing management objectives and should be used.

The purpose of specialized grazing schemes is to improve the carrying capacity of the range. The BLM state reports indicate that the range has not been monitored in a way that establishes the true carrying capacity of the land.[75] In the BLM grazing EISs, current range trends are often inadequately identified. A range inventory should be conducted and accurate baseline data gathered so that trends in range condition can be determined. If a range is in poor condition, but the trend is improving, present management may be justifiably continued and the stocking rate perhaps increased in the future. On the other hand, if a range is in good condition, but the trend is declining, specialized systems may be useful.

In their report to the Council on Environmental Quality, Box, Dwyer, and Wagner note that BLM offices in most states were not implementing basic range management principles: "In many cases there was a tendency to do nothing unless full-scale rest rotation schemes could be implemented."[76] That this tendency prevails is understandable, but disturbing. One might suspect that the agency is holding out so that a better case can be made for its management schemes. If this is true, the agency will probably continue to do so until it is stopped by outside pressures or until it gets its way.

In general, the market system operates through price structures and moves resources to their most highly valued use. The BLM, as a governmental agency, is not confined by the efficiency constraints of the market and it has no reliable price information to guide its behavior in the production of nonmarket goods and amenities.

If BLM lands were transferred to private individuals, there would be direct incentives to increase forage production. However, such action is precluded by the Federal Land Policy and Management Act of 1976, which specifies that public lands will be retained under federal management. In addition, it is likely that private ownership would fail to provide desirable public goods, such as wildlife, watershed protection, and aesthetic quality.

So we must assume that the BLM will continue to administer public lands. If a structure is initiated so that grazing privileges take on the nature of property rights without subsidies, then livestock operators will have incentives to manage their allotments in the most efficient manner and the U.S. Treasury will be provided with a fair return on public forage values. The efficient production of nonmarket values is a more difficult problem; optimal management objectives should be determined in the political arena.

Increasing public knowledge of the possible consequences of BLM policies may help to determine optimal management schemes. Assigning responsibility for management consequences to an agency, although not an easy task, will also help. If, for example, it is widely broadcast that a rest rotation grazing scheme caused sheet erosion and wildlife die-outs on a certain allotment, then the mistake is not so likely to be repeated.

Much is at stake. BLM range lands constitute a large percentage of the total public lands. Those nearest to the public range benefit most directly from it, but people who never see the public lands expect them to be managed so that they, too, will benefit. Public confidence in the ability of agencies to govern benignly is at a low point. An optimum range management decision by the BLM would not only benefit the millions of resource users, it could also improve the morale of the country as a whole.

Notes

1. U.S. Department of Interior, Bureau of Land Management (BLM), *Draft Proposed Wilderness Policy and Review Procedure* (Washington, D.C.: Government Printing Office, 1978).

2. U.S. Department of Interior, Bureau of Land Management, *Final Environmental Impact Statement: Livestock Grazing Management on National Resource Lands* (Washington, D.C.: Government Printing Office, 1974).

3. U.S. Department of Interior, Bureau of Land Management, *Range Condition Report Prepared for the Senate Committee on Appropriations* (Washington, D.C.: Government Printing Office, 1975).

4. The original judgment on the suit called for 212 EISs to be written covering 150 million acres of BLM land. An amendment judgment of April 14, 1978, revised this number to 153 EISs covering about 171 million acres of land under BLM jurisdiction.

5. U.S. Department of Interior, Bureau of Land Management, *Final Environmental Impact Statement: Proposed Domestic Livestock Grazing Program for the Challis Planning Unit* (Washington, D.C.: Government Printing Office, 1976); idem, *Draft Environmental Impact Statement: Uncompahgre Resource Area Grazing Environmental Statement* (Washington, D.C.: Government Printing Office, 1978); idem, *Draft Environmental Impact Statement: The Proposed Rio Puerco Livestock Grazing Management Program* (Washington, D.C.: Government Printing Office, 1978); and idem, *Draft Environmental Impact Statement: San Luis Resource Area Grazing Environmental Statement* (Washington, D.C.: Government Printing Office, 1977).

6. U.S. Department of Interior, *The Public Lands* (Washington, D.C.: Government Printing Office, 1966).

7. As will be discussed later, private property rights generally help to protect the productive capacity of land. A property owner usually benefits most when he protects the value of his property for resale purposes. When property is very cheap, however, and when one is dealing with a resource such as trees where the inventory is high (so that the value of lumber is not expected to increase much with time) and the regeneration period is long, a private landowner may be able to maximize his profits by cutting down all of his trees in the easiest manner and then abandoning his land rather than carefully harvesting timber and replanting seedlings so that the trees can

again be harvested in sixty years or more. This is especially likely when the landowner must pay annual taxes on his land.

8. T. W. Box, D. D. Dwyer, and F. H. Wagner, "The Public Range and Its Management" (Report to the President's Council on Environmental Quality, Utah State University, Logan, 1976).

9. Garrett Hardin, "The Tragedy of the Commons," *Science* 162 (1968):1,244.

10. See Box, Dwyer, and Wagner, "The Public Range."

11. Ibid.

12. For a graphic case history, read Phillip O. Foss, "Battle of Soldier Creek," in *Politics and Grass* (Seattle: University of Washington Press, 1960).

13. Ibid.

14. Ibid.

15. Federation of Rocky Mountain States, "The Federal Land Policy and Management Act of 1976 (Public Law 94-579)" (Briefing paper for the Western Governors' Conference, Denver, Colorado, 1977).

16. Pacific Consultants, "The Forage Resource" (Report produced for the Public Land Law Review Commission, University of Idaho, 1969).

17. U.S. Department of Interior, Bureau of Land Management, "Range Condition Report, National Resource Lands, Colorado," mimeographed (Washington, D.C., 1975).

18. U.S., Congress, Senate, *The Western Range,* 74th cong., 2d sess., Senate doc. 199, 1936.

19. Box, Dwyer, and Wagner, "The Public Range." The reason we might expect private land to be in better shape than BLM lands was touched upon earlier. First, many private ranches were never overgrazed as badly as the public range, since they were not initially a part of the commons situation. Second, ranchers benefit directly from maintaining their land in a healthy, productive state. BLM officials tend to benefit from the political support of their constituents. Due to strong pressures from livestock interests, they have not raised forage fees to market value, and they have not always lowered the number of permitted cattle and sheep within the carrying capacity of the range. The range users, who do not own the range and who are not certain that they will be able to continue using it, also have no real incentive to conserve range quality. So, in a sense, the tragedy of the commons continues on BLM land. Forest Service range land faces some of the same problems. However, the Forest Service began managing its range land in 1905, twenty-nine years earlier than the Grazing Service. Much of the land now administered by the Forest Service was never in as bad a condition and has had a longer time to recover than land administered by the BLM. In addition, Forest Service land is usually at a higher elevation and wetter than BLM land and thus recovers more quickly from abuse.

20. BLM, *Range Condition Report Prepared for the Senate Committee on Appropriations.*

21. U.S. Department of Interior, Bureau of Land Management, "Effects

of Livestock Grazing on Wildlife, Watershed, Recreation and Other Resource Values in Nevada," mimeographed (Washington, D.C., 1974). Since the welfare of bureaucrats tends to increase when the size of the agency is increasing, one would hardly expect an in-house study to reach any other conclusion. For further discussion of this point, see R. McKenzie and G. Tullock, *The New World of Economics* (Homewood, Ill.: Richard D. Irwin, 1975) and W. A. Niskannen, *Bureaucracy and Representative Government* (Chicago: Alden-Atherton, 1971). So a report of this nature must be examined very critically. However, after comparing the budget of the BLM with her sister agency, the Forest Service, and after reading the opinions of respected range scientists (i.e., Box, Dwyer, and Wagner, "The Public Range"), one concludes that there might well be truth to this allegation.

22. Box, Dwyer, and Wagner, "The Public Range."

23. Council on Environmental Quality, *Environmental Quality* (Washington, D.C.: Government Printing Office, 1976), p. 85.

24. K. W. Wiles, "Division of the Range" (Paper given to the Council on Environmental Quality, U.S. Department of Interior, Washington, D.C., 1976).

25. Federation of Rocky Mountain States, "Summary of Issues Related to Grazing on Public Lands in the Rocky Mountain Region," mimeographed (Regional background paper, Denver, Colorado, 1976).

26. BLM, *Range Condition Report Prepared for the Senate Committee on Appropriations.*

27. Public Land Law Review Commission (PLLRC), *One Third of the Nation's Land* (Washington, D.C.: Government Printing Office, 1970).

28. Ibid., p. 105.

29. L. J. Marshall, "Rest Rotation: A New Concept in Range Management," *Western Livestock Journal* (January 1975).

30. Robert E. Steger, "Grazing Systems for Range Care," mimeographed (Albuquerque: New Mexico State University, December 1970).

31. Jared G. Smith, "Forage Conditions of the Prairie Region," *Yearbook of Agriculture* (1895), p. 323.

32. Jared G. Smith, *Grazing Problems in the Southwest and How to Meet Them* (Washington, D.C.: Government Printing Office, 1899), p. 3.

33. A. L. Hormay, *Principles of Rest-Rotation Grazing and Multiple Use Land Management* (Washington, D.C.: Government Printing Office, 1970), pp. 15–16.

34. Council for Agricultural Science and Technology (CAST), "Review of USDA-BLM Draft Environmental Impact Statement Proposed Livestock Grazing Program for the Challis Planning Unit," Report no. 58 (Ames: Iowa State University, 1976), p. 6.

35. A. L. Hormay and M. W. Talbot, *Rest-Rotation Grazing . . . A New Management System for Perennial Bunchgrass Ranges* (Washington, D.C.: Government Printing Office, 1961).

36. BLM, *Uncompahgre Resource Area Grazing Statement,* p. iii.

37. For example, H. F. Heady and J. Bartolome, *The Vale Rangeland Rehabilitation Program: The Desert Repaired in Southeastern Oregon* (Portland: U.S. Department of Agriculture, Pacific Northwest Forest and Range Experiment Station, 1977).

38. See H. F. Heady, "Continuous vs. Specialized Grazing Systems: A Review and Application to the California Annual Type," *Journal of Range Management* 14 (1961):182–93; and D. N. Hyder and W. A. Sawyer, "Rotation-Deferred Grazing as Compared to Season-Long Grazing on Sagebrush-Bunchgrass Ranges in Oregon," *Journal of Range Management* 4(1951):30–34.

39. W. C. Hickey, *A Discussion of Grazing Management Systems and Some Pertinent Literature (Abstracts and Excerpts, 1895–1966)* (Denver: U.S. Department of Agriculture, 1969).

40. Box, Dwyer, and Wagner, "The Public Range," p. 39.

41. R. D. Ratliff, J. N. Reppert, and J. R. McConnen, *Rest-Rotation Grazing at Harvey Valley . . . Range Health, Cattle Gains, Costs* (Berkeley, Calif.: U.S. Department of Agriculture, Pacific Southwest Forest and Range Experiment Station, 1972).

42. W. C. Cook, *Effects of Season and Intensity of Use on Desert Vegetation,* Agricultural Experiment Station Bulletin 483 (Logan: Utah State University, 1971).

43. See, for example, the EIS released in 1978 for the Hot Desert Resource Area in Utah.

44. R. S. Driscoll, *Managing Public Rangelands: Effective Livestock Grazing Practices and Systems for National Forests and National Grasslands* (Washington, D.C.: Government Printing Office, 1967).

45. T. N. Shiflet, and H. F. Heady, *Specialized Grazing Systems: Their Place in Range Management* (Washington, D.C.: Government Printing Office, 1971).

46. "Rest-Rotation Plan: Panacea or Problem?" *High Country News,* January 14, 1977.

47. A. L. Hormay and A. B. Evanko, "Rest-Rotation Grazing: A Management System for Bunchgrass Ranges" (U.S. Department of Agriculture, California Forest and Range Experiment Station, 1958); W. M. Johnson, "Rotation, Rest-Rotation, and Season-Long Grazing on a Mountain Range in Wyoming" (U.S. Department of Agriculture, Rocky Mountain Forest and Range Experiment Station, 1965); L. Rader, "Grazing Management: Payson Perennial Grass During Drought" (U.S. Department of Agriculture, Pacific Forest and Range Experiment Station, 1961); R. D. Ratliff and L. Rader, "Drought Hurts Less with Rest-Rotation Management" (U.S. Department of Agriculture, Southwestern Forest and Range Experiment Station, 1962); and E. J. Woolfolke, "Rest-Rotation Management Minimizes Effects of Drought" (U.S. Department of Agriculture, Pacific Southwestern Forest and Range Experiment Station, 1960).

48. J. P. Workman and M. Nazir, *An Analysis of Bureau of Land Management Grazing Systems in the Intermountain Area* (Logan: Utah State University,

1975); and D. G. Rozell, C. T. K. Ching, and C. E. Hancock, *Economic Returns of Rest Rotation Grazing to the Bureau of Land Management* (Reno: University of Nevada, 1973).

49. G. F. Gifford and R. H. Hawkins, "Grazing Systems and Watershed Management: A Look at the Record," *Journal of Soil and Water Conservation* 31 (1976):282.

50. BLM, *San Luis Resource Area Grazing Statement.*

51. G. F. Gifford, R. H. Hawkins, and J. S. Williams, "Hydrologic Impacts of Livestock Grazing on National Resource Lands in the San Luis Valley" (Completion report to the Bureau of Land Management, Utah State University Foundation, Logan, 1975).

52. Even if a pasture has recovered somewhat, the erosion rate is likely to be higher with the occurrence of a high intensity storm. See BLM, *San Luis Resource Area Grazing Statement.*

53. BLM, *Uncompahgre Resource Area Grazing Statement,* p. 3-59.

54. Ibid., p. 3-60.

55. BLM, *Livestock Grazing Program for Challis,* p. 5-3.

56. A. J. Cordone and D. W. Kelley, "The Influences of Inorganic Sediment on the Aquatic Life of Streams," *California Fish and Game* 47 (1961):189-228.

57. BLM, *Uncompahgre Resource Area Grazing Statement,* p. 3-64.

58. B. Shanks, "Survey of Deer Hunters in Northern California" (Logan: Utah State University, 1976).

59. BLM, *Range Condition Report Prepared for the Senate Committee on Appropriations.*

60. Rozell, Ching, and Hancock, *Economic Returns of Rest Rotation Grazing.*

61. Workman and Nazir, *BLM Grazing Systems.*

62. D. B. Gardner, "Transfer Restrictions and Misallocation in Grazing Public Range," *Journal of Farm Economics* 44 (1962):109-20.

63. U.S. Department of Interior, Bureau of Land Management, *Allotment Management Plan (AMP). Economic Analysis Instruction,* Memorandum no. 76 (Washington, D.C.: Government Printing Office, 1976), p. I-16.

64. BLM, *Wilderness Policy and Review Procedure,* p. 3-3.

65. PLLRC, *One Third of the Nation's Land,* p. aa2.

66. For a more completely developed model of a similar proposal, see John A. Baden and Richard L. Stroup, "Private Rights, Public Choices, and the Management of National Forests," *Western Wildlands,* Autumn 1975, pp. 5-13.

67. Gardner, "Transfer Restrictions."

68. D. B. Gardner, "A Proposal to Reduce Misallocation of Livestock Grazing Permits," *Journal of Farm Economics* 45 (1963):109-20.

69. Ibid.

70. CAST, "Review of USDA-BLM Draft EIS."

71. Shiflet and Heady, *Specialized Grazing Systems,* p. 10.

72. Ibid.

73. W. C. Cook, "Summary," *Review of Draft Environmental Impact Statement* (Fort Collins: Colorado State University, 1976), p. 8.

74. McKenzie and Tullock, *The New World of Economics.*

75. Box, Dwyer, and Wagner, "The Public Range."

76. Ibid., p. 21.

Chained to the Bottom

by Ronald M. Lanner

Picture yourself driving across one of those broad desert valleys in the West under a brassy summer sun. You look ahead with anticipation as the road climbs gradually into a range of mountains with broad slopes studded with scattered, spreading trees. Soon you are cooling yourself in the shade, contentedly nibbling cheese and crackers and quaffing cold beer. Just as you congratulate yourself on your good luck and foresight, you hear the roar of diesel engines and the clacking of caterpillar tracks. Comestibles in hand, you leap to your feet to confront a pair of enormous Forest Service bulldozers coming your way in twin clouds of dust. They drag between them a great gleaming steel chain that smites the trees in its path and wrenches the very boulders from their earthly beds. Your tree's turn is coming up soon, so you retreat to the highway. Being endowed with a healthy curiosity, you wait around to ask the tractor operators why they make these great swaths and clearings across the face of the land. You learn that their tractors are veritable Engines of the Public Good, enhancing the flow of water, reducing erosion, augmenting our food supply, succoring animals in distress, protecting trees from insects and disease, beautifying the landscape, safeguarding the nation's archaeological heritage, and preserving the rural health and economy. Reverently you watch the great dust cloud dissipate over the desert valley as you resolve to write your congressman to urge higher appropriations for the Great Work you have been privileged to witness.

This article is adapted from Ronald M. Lanner, "The Eradication of Pinyon-Juniper Woodlands: Has the Program a Legitimate Purpose?" *Western Wildlands,* Spring 1977, pp. 12–17. These issues are discussed further in Ronald M. Lanner, *The Piñon Pine—A Natural and Cultural History* (Reno: University of Nevada, forthcoming). Used by permission.

This scenario is admittedly somewhat fanciful, but given an impressionable observer and a supersalesman espousing some of the extravagant claims that have been made on behalf of "chaining" programs, it could almost happen that way. Agencies like the Forest Service *do* push over little trees, by the scores of thousands of acres; and their agents and assigns *do* justify such projects in terms suggesting that a deity lies somewhere close by the regional forester's right hand.

In this paper, I will identify the costs and benefits of this federal program. I will also go to some trouble to point out that the basic justification for the program is fallacious, because a malfunctioning information system seems to be a prerequisite for its continued life. But first, let us look at the geographic and technological context of the chaining program.

The Woodlands

At middle elevations in the Southwest and Great Basin there are about 60 million acres of woodland dominated by small pines (piñons) or junipers. At their lower edges these woodlands are savannalike and are composed mainly of juniper. Canopy density increases with elevation, and piñon becomes dominant. The upper edge of the woodland usually merges with coniferous forest. The elevation ranges from 5,000 to 8,000 feet, and precipitation is about ten to twenty inches.

The woodlands of western Utah and Nevada—the area we will be concerned with—usually occupy the slopes of mountains and the floors of high valleys in the Great Basin physiographic province.[1] For many centuries, these woodlands were vital sources of food, fuel, and fiber for early Indian societies on the Plateau and in the Great Basin. They were the habitat of most prehistoric peoples of the Southwest, as shown by the heavy concentration of archaeological sites and artifacts. After white settlement and until the decline of the fuel wood market and the advent of steel fence posts, the woodlands were heavily cut. Since settlement, they have been grazed by cattle, often to the point of severe abuse. The woodland forms the largest forest type in the Southwest and Great Basin areas.

The vast bulk of woodland is in public—overwhelmingly federal—ownership. Utah has about 11,700,000 acres of woodland, of which 754,000 are in national forest and administered by the U.S. Forest Service; Nevada has 5,900,000 acres, of which 1,900,000 are in national forests. Most of the remainder in both states is administered by the Bureau of Land Management (BLM), successor to the U.S. Grazing Service. Woodlands in the national forests of Utah and Nevada are in the Forest Service's Intermountain Region, which is governed by the regional forester in Ogden, Utah. The BLM is organized by state jurisdictions; its Utah and Nevada woodlands are managed under the aegis of state directors in Salt Lake City and Reno. For the sake of brevity, I will refer to these units as the Forest Service, and the Utah and Nevada BLM.

The Treatment

Chaining is the practice of attaching the ends of an anchor chain to two crawler tractors, which then drive parallel to each other dragging the chain through the stand. The chain may be as long as 600 feet with links weighing up to 90 pounds each.[2] Often, two passes must be made, because small trees tend to ride under the chain, thus resisting efforts to uproot them, and because some trees need to be uprooted on both sides before they will fall. Debris may be left in place or windrowed. The area is then seeded with forage plants, especially the Central Asian exotic, crested wheatgrass. Smaller areas are cleared by burning, spraying, hand cutting, or bulldozing. In its Southwestern region, the Forest Service uses a recent innovation, the eighty-ton "Tree Crusher." As foresters drive this large machine through woodlands, the blades on its electrically driven wheels chop trees into firewood-sized chunks, but it cannot be used in rough or rocky terrain.

Between 1960 and 1972, the BLM chained more than 250,000 acres in Utah and 43,000 acres in Nevada.[3] Over 77,000 acres have been chained in Utah's national forests, and almost 6,000 acres in the Humboldt Forest in Nevada, according to information received from Forest Service officials. No data have been forthcoming from the Toiyabe Forest in Nevada. Most chaining has actually been done in Arizona and New Mexico, and, when those states are included, the total area cleared between 1950 and 1964 rises to about 3 million acres.

As chaining got underway, there were numerous protests from concerned citizens, but no serious organized opposition appeared because the great environmental movement had not yet been born. Still, the heavy-handedness of the technique, its drastic effects on the landscape, and its apparent lack of benefit for anybody except holders of cattle leases on public land have subjected the program to skepticism for two decades.

Neither the Forest Service nor the BLM went to any trouble to publicize details of their chaining programs until the National Environmental Policy Act required that environmental impact statements (EISs) be prepared. The Forest Service complied by preparing nearly identical EISs for their proposed chaining in Utah and Nevada. The BLM does not yet appear to have prepared such statements.

Why Chain?

The effectiveness of an action can be considered separately from its rationale; or as we often observe, the right (or wrong) thing can be done for the wrong (or right) reasons. Before looking at the results of chaining, let us consider its rationale.

According to the Forest Service EISs, chaining is a plant *control* program. In fact, the Forest Service refers to the 110,000 acres in Utah and 287,000 acres in Nevada where chaining is planned (at the rate of 10,000 to 13,000 acres per year) as being "suitable for rehabilitation." Rehabilitation means restoration to a former state. Thus, the basic assumption is that chaining must be applied to lands that were historically grassland or shrubland, but where the woodland has *invaded*.

Here are but a few of the statements in the Utah EIS expressing alarm over the invasive behavior attributed to the woodland:

> In many of these woodland plant communities the value of land . . . has been seriously reduced, as the stands have thickened and increased their area of occupation.[4]

> During the past hundred years or so these woodland areas have been spreading.[5]

> . . . as the stands continue to expand and trees mature and completely occupy the type.[6]

The EIS unquestionably accepts the idea that the woodland is aggressively moving into areas that were formerly grassland or shrubland. There is puzzlement over whether this is due to overgrazing, fire protection, or even climatic change; but the concept of massive region-wide invasion is not questioned.

This is most important because it gives the chaining program a sense of mission. There is even an aura of old melodrama, with the private rancher as victim, the woodlands as villain, and the agency the hero with a bulldozer on each hip. Yet, the validity of the invasion hypothesis remains undemonstrated and open to serious question.

To validate the invasion hypothesis two things are essential. First, it must be shown that a given area was grassland or shrubland just prior to settlement and, second, that it later became wooded. It is not enough to show "before and after" photo pairs in which the early photo records the vegetation *during* the period of settlement. Such pictures may show extensive open areas in the hills around villages, with woodland present only on the higher mountains;[7] but they tell nothing about the vegetation on the lower slopes *before* white settlement. They may simply be recording deforested slopes stripped of lumber, posts, fuel, and other necessities of early rural life. Yet photos of such limited value have been offered as primary evidence of woodland invasion.

Until the extent of the early woodland deforestation is established, it will not be possible to determine whether young stands of piñon and juniper (up to a hundred or so years of age) constitute an invasion into new territory, or merely the reestablishment of woodland on its former sites. This is especially the case in Nevada, where historic evidence establishes that widespread clear-cutting of woodlands occurred over a period of several decades. The significance of man's prior activities on the makeup of today's piñon-juniper woodland can be seen by looking at Eureka, an old mining center in central Nevada.

Ore smelting began in Eureka in the early 1870s. Thirteen furnaces were operating by 1874, and there were sixteen by 1878. In 1878, the Nevada Surveyor-General reported that the smelters consumed 16,000 bushels of piñon charcoal daily. Since a typical piñon yield was ten cords per acre, and a cord made thirty bushels of charcoal, the furnaces of Eureka devoured over fifty acres of woodland a day.[8] By 1873, all the piñon within ten miles of Eureka had been cut,[9] and by 1878, according to the Nevada state mineralogist, the denuded area extended for a radius of fifty miles.[10] A fifty-mile radius of Eureka approaches to within a few miles of Ely to the east and Austin to the west. Both of these towns were also important mining centers, whose demands for charcoal and other woodland products may have rivaled those of Eureka.

A great deal of careful research must be done to clarify the effects of "coaling" on Nevada's forests, but generalizations about woodland invasion cannot safely ignore those effects. On the whole, I think the magnitude of invasion has been vastly exaggerated, largely because the scope of man's past activities has been ignored. This applies to Utah, Arizona, and New Mexico as well as Nevada.

Curiously, the Forest Service does not *act* as though it really puts much credibility in the invasion hypothesis at all. If it were indeed rehabilitating invaded lands, then it would chain only those areas where invasion was known to have occurred. But no inventory of woodlands has been made to locate such areas, and no map exists that pinpoints invasions. The 397,100 acres slated for chaining in Utah and Nevada were not selected because they were invaded but because they were convenient. So while the Forest Service justifies its chaining program as a rehabilitation measure to control an invading forest, it routinely applies the method to old stands as well as young ones and plans to continue doing so. As the EISs actually say, "Often the stands selected are old and mature, and very thick."[11]

This all seems very confusing until we explore the Forest Service's likely motives for chaining. First it must be understood that the Forest Service has always had problems managing the woodland *as a woodland*. It is widely dispersed and relatively unproductive. The harvesting of wood products is often economically infeasible, and the products that are removed (firewood, posts, pine nuts) tend to be consumed locally and have little impact on the regional market economy. Since settlement, the unsupervised removal of tree products and heavy grazing have resulted in severe degradation of the woodlands, further reducing their ability to produce more such products. The Forest Service responded by labeling the woodlands "noncommercial forest" and began managing them as rangelands. Thus, active, on-the-ground management passed from frustrated timber-oriented foresters to range managers whose professional objective is the production of red meat. Trees are more of a hindrance than a resource to range managers, and chaining is an attractive method of removing them. Identification with client groups of stockmen probably further convinced range managers of the value of chaining.

Perhaps of equal importance is the strongly felt need to get out there and do something to make ideal land productive. Here is a revealing statement by Forest Service EIS writers considering the alternative to "no action":

The Forest Service is charged by law to manage National Forests . . . on a sustained-yield basis. The alternative of taking no action does not meet the objectives of sustained-yield management, when environmentally sound methods are available for improving this phase of resource management.[12]

Conversations with Forest Service range managers have convinced me that they also favor chaining because it is a secure budgetary item and because it makes friends for the agency among influential stockmen. It has also been pointed out to me that range lessees can be better made to comply with Forest Service requirements if the prospect of chaining federal lands is dangled before them.

For all of these reasons chaining has enjoyed strong Forest Service support since the early 1950s. Twenty years later, when a new law mandated the preparation of environmental statements, the Forest Service prepared advocacy documents rather than balanced weighings of objective evidence. Given the ideology and bias of the agency, it is doubtful that they could have done otherwise.

Impacts on Selected Resources

In this section we will identify the more important impacts, both positive and negative, on the major resources of the piñon-juniper woodland.

Livestock Forage

There is no doubt that removal of tree cover provides opportunities for increasing forage production. Dwyer indicates that chaining provided significantly larger crops of forage than grew in the woodland, ranging from no change to a 600 percent increase.[13] The forage increase is often directly related to density of the stand of trees removed. No negative impacts of chaining on livestock forage have come to my attention.

Watershed Values

In the Forest Service's view, stands of trees shade understory plants which then die, leaving the soil subject to erosion.[14] It is, therefore, their position that chaining *reduces* erosion despite the mechanical soil disturbance it causes, assuming, of course, that a dense stand of grass quickly becomes established. The Forest Service also believes that removal of tree cover will reduce on-site water consumption, allowing more runoff to feed into river systems.

Neither of these potential benefits has been supported by research. Gifford has reviewed the literature on water quality, runoff, infiltration, and interception and has concluded that no improvement in these aspects of the watershed can be reliably expected as a result of chaining.[15] A study by Forest Service scientists also shows that chaining has no significant effect on water quality or sediment yields.[16] The same study concludes that "mechanical methods of piñon-juniper removal are not likely to increase water yield." Thus, there are no established watershed benefits that result from chaining.

Wildlife Habitat

The Forest Service considers wildlife a major beneficiary of woodland eradication. Emphasis is placed on the mule deer herd, because of its importance as a recreational resource. During the winter, mule deer feed heavily on piñon-juniper stands, and the Forest Service's position is that openings created by chaining provide needed forage. The results of many years of research, however, do not support this optimism. A recent review found that "a summary conclusion for the several million acres of treated P-J is one of *no overall impact—either positive or negative.*"[17] Similar conclusions have been reached by Forest Service scientists in Arizona and New Mexico.[18]

An important reason why the mule deer do not use extensive chained areas is their hesitancy to expose themselves in large openings, and many chainings are hundreds or even thousands of acres in size. To enhance their value for deer, chainings must be small in size. In other words, if chaining is to improve deer habitat, it should be done by "punching" small holes in the woodland matrix, not by leaving wooded "islands" surrounded by large cleared areas.[19] But chaining done in this manner compromises livestock benefits and raises costs.

The neutral response of deer to most chaining is important. The Forest Service holds strongly that a positive benefit accrues in winter feed availability. It is implied that this easily accessible feed increases herd size and ultimately augments the hunter's take, bringing sizable economic benefits. But with the repeated findings of neutrality, this argument becomes a house of cards.

According to the Forest Service, there are about 50 species of fish, 66 reptiles and amphibians, 75 mammals, and 140 birds in or around the piñon-juniper woodlands.[20] Terrel and Spillett point out that "we remain in near total ignorance of the impact of P-J conversion on the hundreds of vertebrate species influenced by this large ecosystem."[21] Such ignorance prohibits a comprehensive listing of animals harmed by chaining. However, a lack of data did not inhibit Forest Service EIS writers from concluding that chaining would benefit owls, hawks, and eagles by providing openings in which to find their prey! Rarely has a governmental agency shown such solicitous regard for its feathered wards.

Archaeological Antiquities

Most Great Basin archaeological sites are found in the piñon-juniper woodland, and chaining has the potential of destroying nine-tenths of them.[22] But a silver lining in this dark cloud is plainly evident in the EISs. "It is possible," we are told, "to visualize the results of an archaeological inventory and other steps taken to protect the resource from chaining impacts as being a favorable impact. The addition of new archaeological data . . . and the economic benefits to local archaeological institutions would be significant."

But the overall destructiveness of driving bulldozers, dragging chains, and uprooting trees is obvious. According to research done by none other than a Forest Service archaeologist, chaining "may totally eliminate any possibility of conducting studies of surface cultural patterns. If such is the case, then adverse site impacts from chaining might more accurately be estimated as nearly 100 percent."[23]

A fascinating aspect of the chaining program is the number of official lawbreakers it may already have created. According to the Antiquities Act of 1906, "Any person who shall appropriate, excavate, injure or destroy . . . any object of antiquity, situated on lands owned or controlled by the Government of the United States, without the permission of the Secretary of the Department of the Government having jurisdiction over the lands on which said antiquities are situated, shall, upon conviction, be fined in a sum of not more than five hundred dollars or be imprisoned for a period of not more than ninety days."

Miscellaneous Impact

The Forest Service claims that several rather novel benefits result from chaining.[24] For example, chaining will increase plant diversity, which is seen as an improvement of the ecosystem. The concern for a lack of diversity (too few plant species) in the woodland is at odds with policy elsewhere and creates ad hoc policy completely at odds with Forest Service timber programs. For example, the Forest Service plants huge acreages of pure conifer plantations throughout its lands: ponderosa pine in Idaho and California, Douglas fir in the northwest, lodgepole pine in the Rockies. In fact, the Forest Service has vigorously defended such monocultures on both economic and biological grounds against environmentalists who attack them for their lack of diversity. When we disregard these inconsistent policy objectives, the argument still fails to be convincing.

According to the Forest Service, the woodland contains twenty-two common shrub species, fourteen grasses, and seventeen forbs.[25] Yet, the vast proportion of chained areas is seeded with pure stands of crested wheatgrass, an Asian import. Even if it were beneficial to increase species diversity in this natural ecosystem, and there is no evidence that this is the case, seeding pure crested wheatgrass would not accomplish that purpose.

The Forest Service also claims "better buffer potential against pests," a vague claim for which no evidence is offered.[26] Woodland is apparently seen as a vegetational analog of the Vietnamese village that had to be destroyed in order that it might be saved. Actually, the woodland is relatively free of pest epidemics. More susceptible forests elsewhere under Forest Service administration are spared the indignity of the dozer and chain, despite the presence of pests. This casually derived Forest Service policy will surely not be extended to any other forests where pests are a threat.

A locally important adverse impact of chaining is that it deprives people of an important food. This is especially true for the Shoshone and Paiute Indians of Nevada, some of whom are still dependent on their traditional winter food, the pine nut. Some families collect three hundred pounds of pine nuts each fall and subsist on little else until spring. The collection and preparation of this nutritious food are important elements in the traditions and lifestyle of these Indians. In Nevada, the BLM has eternally tarnished its image by chaining some of the best and most accessible piñon stands. Many of the Indians who go pine-nutting in the fall are old; they resent *all* chaining, but especially the chaining of low elevation stands that are easily reached. They see their own food supply being destroyed to accommodate white men's cattle, and their bitterness has come to fruition in legal action against the BLM. At this writing, the lawsuit brought in federal court by the Indians of Battle Mountain, Nevada, has not been resolved, but the BLM has managed to alienate a significant community.

Health surely must be considered in any public program, and the EISs show ample humanitarian concern. How does chaining affect human health? Life has its hazards, of course, but on balance the program stands vindicated. "The only known hazard to human health," we learn, "is the possibility of one of the workmen getting hurt during the process of chaining." But the "improved habitat conditions" that follow chaining "may affect a person's well-being through increased participation in sightseeing, hunting, [and] photography." Most of the alternatives to chaining (prescribed burning, single tree selection, tree crushing) are not known to affect human health one way or the other, but if *nothing* is done, the consequences may be dire. Admittedly, "nonremoval of the trees may improve someone's mental health or attitude toward the environment." But on the other hand, "without [tree removal] the trees will continue to spread, thus causing a loss of valuable forage needed for wildlife and livestock needs. This loss causes a loss of income to families who depend on livestock and hunting as a source of their income. This could very well affect their human health."

The Economics of Chaining

Benefit-cost analyses of chaining were not made until the 1970s, during the third decade of the program. Nor are they universally used in justifying chaining programs. For example, the final environmental statements for mechanical treatment of vegetation in Arizona and New Mexico make no attempt whatever to compare project costs and benefits.[27]

Published benefit-cost studies include one prepared by the authors of the Forest Service environmental statements for Utah and Nevada,[28] one prepared by Kienast for the BLM,[29] and one made by Forest Service researchers in Arizona.[30] These have been reviewed by Workman and Kienast.[31] All of the analyses include the value of increased forage as a market-priced benefit. The benefit-cost analyses that the Forest Service included in the EISs[32] also include monetary benefits from reduced soil erosion and from the economic value of enhanced mule deer hunting. However, as shown above, there is no evidence that chaining either reduces erosion or increases the hunt, so inclusion of these illusory benefits is unwarranted.

The analysis that includes these items assumes that chained clearings will remain open for fifty years and that grazing will be done during forty-three of those years. Costs of clearing are discounted at 10 percent. It arrives at a benefit-cost ratio of 1.08:1. In other words, for every public dollar invested in the program, $1.08 will accrue to the economy. Workman and Kienast recomputed this analysis to delete the unjustified erosion and hunting benefits and to make the analysis comparable to other studies.[33] They used a discount rate of only 7 percent and a fifty-year discount factor of 13.8 and arrived at a benefit-cost ratio of only 0.86:1.

Actually, the assumed fifty-year life of a chained area is sheer speculation, for none has been in existence for more than about twenty years. Many chained areas have required additional treatment to remove newly established tree cover within fifteen to twenty years, and total clearing costs as high as $70 per acre have been reported (almost four times the cost used in the above analysis).[34] It is astonishing that, despite access to hundreds of thousands of acres of artifical clearings, neither the Forest Service nor the BLM has accurate data on the "life expectancy" of such clearings. Apparently, there has been no serious attempt to learn how many years will elapse before forage production is significantly reduced by reinvasion of woodland trees. The one study cited in the Forest Service EISs showed that released seedlings began to dominate the site within twelve to fifteen years.[35] Thus, no data exist to justify use of a fifty-year life for clearings in computing the economic benefits of chaining. The above analysis would show a much less favorable return if a clearing life of only twenty years were used.

Kienast's analysis assumed a twenty-five-year project life and a 7 percent discount rate. His benefit-cost ratios varied widely from 0.30:1 to 1.32:1. The more favorable ratio resulted when uprooted vegetation was allowed to remain in place, while the most unfavorable situation resulted from the increased cost of pushing the debris into piles.[36] This suggests that the cost of operating tractors—which has escalated with fuel prices since these studies were made—can be a critical factor in the economics of chaining.

Forest Service researchers used an infinite project life and a 7 percent discount rate. They concluded that, under 1972 economic conditions, "the more successful conversion projects just about break even from a benefit-cost standpoint."[37]

In view of these findings, it is obvious that in terms of livestock forage—the only resource that is agreed by all to be enhanced by chaining—a break-even situation is about all that can be hoped for.

Avoiding Pitfalls

If the current chaining program has been a mistake—as I believe is the case—how could it have been avoided? There are two elements to a more rational approach to woodland management.

Improving the Information System

The chaining program began in an informational vacuum. There was no way of knowing at the outset whether it would pay its way or what the magnitude of its future negative impacts would be. Yet it developed rapidly and reached large scale application within a few years. It was only many years later that its diseconomies were discovered. All this could have been avoided if the Forest Service had done two things well within its capabilities. It could have established a series of pilot studies and monitored them over a period of years, and it could have called in expert assistance from its own research branch, in this case the Intermountain Forest and Range Experiment Station. These measures would have gone far to provide managers with reliable and relevant information, including cost-effectiveness, *before* embarking on a continuing program. Further improvement of the information system would come about if environmental statements were written by disinterested professionals with analytical skills, not by staff directly involved with the program being evaluated. Again, the Forest Service has a pool of such talent in its experiment stations, and there are a number of private consultants who can prepare statements on a contract basis. The currently active chaining EISs are beset with biases, misinformation, and preconceived notions that vindicate the program that they are supposed to evaluate.

Seeing the Woodland as a Forest

The present chaining program would perhaps not have come about if the woodland were regarded as a forest producing simultaneous yields of wood products, vegetable protein (pine nuts), wildlife, recreational opportunities, *and* livestock forage instead of an expensive grass. A forest can be actively managed for such multiple uses, with the mix of products and the level of investment determined by local economic and biological conditions.

Apparently, both of these elements have now come together in the Forest Service's Southwestern Region, comprising the national forests of Arizona and New Mexico. A recent management directive, tentative but pointing in the right direction, is currently being adopted there. Its major components include detailed inventory of the woodland resource, use of cost-effectiveness as a management criterion, and multiple use management based on land capability. Areas that have been chained or otherwise converted to grassland will be reexamined to determine their future role. Certified Forest Service silviculturists will be "deeply involved" in woodland management.[38] Strict procedures will safeguard archaeological values.

These actions are being taken not in the absence of information, but during a pronounced surge of research activity by the Rocky Mountain Forest and Range Experiment Station. It is hoped that the Southwestern woodlands will eventually regain their original productivity, and those of the Intermountain Region and under BLM control will be placed under similar rational management.

Notes

1. Details of geographic and ecological distribution of these woodlands have been published elsewhere. See Ronald M. Lanner, "Pinyon Pines and Junipers of the Southwestern Woodlands," in *The Pinyon-Juniper Ecosystem: A Symposium*, ed. Gerald F. Gifford and Frank E. Busby (Logan: Utah State University, 1975), pp. 1–17; Paul T. Tuller and James E. Clark, "Autecology of Pinyon-Juniper Species of the Great Basin and Colorado Plateau," in ibid., pp. 27–40; and Neil E. West, Kenneth H. Rea, and Robin J. Tausch, "Basic Synecological Relationships in Juniper-Pinyon Woodlands," in ibid., pp. 41–53.

2. Richard S. Aro, "Pinyon-Juniper Woodland Manipulation with Mechanical Methods," in *The Pinyon-Juniper Ecosystem*, pp. 67–75.

3. Ibid.

4. U.S. Department of Agriculture, Forest Service, "Pinyon-Juniper Chaining Programs on National Forest Lands in the State of Utah" (Ogden, Utah, Intermountain Region, 1973), p. 2.

5. Ibid., p. 5.

6. Ibid., p. 26.

7. Ibid.; idem, "Pinyon-Juniper Chaining Programs on National Forest Lands in the State of Nevada" (Ogden, Utah, Intermountain Region, 1974); and West, "Basic Synecological Relationships," in *The Pinyon-Juniper Ecosystem*.

8. Phillip E. Earl, "Nevada's Italian War," *Nevada Historical Society Quarterly* 12 (1969):47–87.

9. Rossiter W. Raymond, *Silver and Gold: An Account of the Mining and Metallurgical Industry of the United States* (New York: J. B. Ford, 1873).

10. Earl, "Nevada's Italian War."

11. Forest Service, "Pinyon-Juniper Chaining Programs in Utah"; idem, "Pinyon-Juniper Chaining Programs in Nevada."

12. Forest Service, "Pinyon-Juniper Chaining Programs in Utah."

13. Don D. Dwyer, "Response of Livestock Forage to Manipulation of the Pinyon-Juniper Ecosystem," in *The Pinyon-Juniper Ecosystem*, pp. 97–103.

14. Forest Service, "Pinyon-Juniper Chaining Programs in Utah"; idem, "Pinyon-Juniper Chaining Programs in Nevada."

15. Gerald F. Gifford, "Impacts of Pinyon-Juniper Manipulation on Watershed Values," in *The Pinyon-Juniper Ecosystem,* pp. 127–41.

16. Warren P. Clary et al., "Effects of Pinyon-Juniper Removal on Natural Resource Products and Use in Arizona," Research Paper RM–120 (U.S. Department of Agriculture, Forest Service, 1974).

17. Ted Terrel and J. Juan Spillett, "Pinyon-Juniper Conversion: Its Impact on Mule Deer and Other Wildlife," in *The Pinyon-Juniper Ecosystem,* pp. 105–19. Emphasis added.

18. Clary et al., "Effects of Pinyon-Juniper Removal"; and Henry L. Short, Wain Evans, and Erwin L. Boeker, "The Use of Natural and Modified Pinyon Pine-Juniper Woodlands by Deer and Elk," *Journal of Wildlife Management* 41 (1977):543–50.

19. Terrel and Spillett, "Pinyon-Juniper Conversion."

20. Forest Service, "Pinyon-Juniper Chaining Programs in Utah."

21. Terrel and Spillett, "Pinyon-Juniper Conversion."

22. Forest Service, "Pinyon-Juniper Chaining Programs in Utah."

23. E. I. DeBloois, D. F. Green, and H. G. Wylie, "A Test of the Impact of Pinyon-Juniper Chaining on Archaeological Sites," in *The Pinyon-Juniper Ecosystem,* pp. 153–61.

24. Forest Service, "Pinyon-Juniper Chaining Programs in Utah"; idem, "Pinyon-Juniper Chaining Programs in Nevada."

25. Ibid.

26. Ibid.

27. U.S. Department of Agriculture, Forest Service, "A Proposal for Vegetation Control by Mechanical Treatment in the State of Arizona: Final Environmental Statement" (Albuquerque, New Mexico, Southwestern Region, 1973); and idem, "Final Environmental Statement of a Proposal for Vegetation Control with Mechanical Treatment in the State of New Mexico," (Albuquerque, New Mexico, Southwestern Region, 1973).

28. Forest Service, "Pinyon-Juniper Chaining Programs in Utah"; idem, "Pinyon-Juniper Chaining Programs in Nevada."

29. C. R. Kienast, "Final Report to Bureau of Land Management," Utah Agricultural Experiment Station Project 728, Obj. 5 (Logan, Utah, 1974).

30. Clary et al., "Effects of Pinyon-Juniper Removal."

31. John P. Workman and Charles R. Kienast, "Pinyon-Juniper Manipulation: Some Socio-Economic Considerations," in *The Pinyon-Juniper Ecosystem,* pp. 163–76.

32. Forest Service, "Pinyon-Juniper Chaining Programs in Utah"; idem, "Pinyon-Juniper Chaining Programs in Nevada."

33. Workman and Kienast, "Pinyon-Juniper Manipulation."

34. Aro, "Pinyon-Juniper Woodland Manipulation."

35. R. J. Tausch, "Plant Succession and Mule Deer Utilization on Pinyon-Juniper Chainings in Nevada" (M.S. thesis, University of Nevada, Reno, 1972).

36. Kienast, "Final Report."

37. Clary et al., "Effects of Pinyon-Juniper Removal."

38. Personal communication from M. M. Johannesen, 1977.

An Institutional Dinosaur with an Ace: Or, How to Piddle Away Public Timber Wealth and Foul the Environment in the Process

by Barney Dowdle

One can easily imagine the derisive comments that would greet a proposal to turn over the management of $42 billion worth of resources to an organization that:

>   1. did not recognize opportunity costs, i.e., alternative ways to employ resources that might be more profitable, efficient, and socially desirable
>   2. did not recognize that efficient resource use requires that superior resources be exploited before inferior ones
>   3. used a procedure for evaluating investment alternatives that gives uneconomical investments the appearance of being highly lucrative.

We do, in fact, have this kind of an organization in the United States today—the Forest Service of the U.S. Department of Agriculture. The Forest Service is responsible for managing the system of national forests in the United States.[1] In addition to ignoring the opportunity costs of the resources it manages, the Forest Service also runs a sizable cash flow deficit. In recent years, Forest Service expenditures have exceeded receipts obtained mostly from timber sales by nearly $500 million per year.

How did we get an organization like this? And given that we have it, why do we keep it? If it is being managed inefficiently, why is there no social demand for reform?[2] This paper is an attempt to address these questions. To use a cliché appropriate to the occasion, the argument emphasizes "looking at the forest rather than the trees."

Most professional foresters are preoccupied with problems that arise within the framework of existing institutional and organizational arrangements. Rarely do they take a critical view of the arrangements themselves and ask if perhaps they might be the source of the problem. There are reasons for this, of course. Agencies that manage public forests are the source of considerable largesse. The old adage, "don't bite the hand that feeds you," is a powerful incentive encouraging forestry researchers to try to make existing arrangements work. To suggest that they are wrong is to court the displeasure of those who dispense sizable amounts of research funds.

An additional reason that the present institutional arrangements persist is that special interest groups that benefit from them have been able to delude the general public into believing that the public interest is being served. The general public seldom takes the trouble to evaluate existing arrangements until a crisis occurs.

The so-called sustained yield-even flow theory that underlies public forest management guidelines is an anachronism—an institutional dinosaur, as the title of this paper suggests. These guidelines are highly wasteful of both public timber wealth and taxpayers' dollars. In addition, environmental impacts are unnecessarily high—not a surprising result given that public forest resources are managed without proper regard for costs.

Background

A popular myth that has flourished in the United States for over three-quarters of a century is that private ownership of timber resources and the marketplace are incompatible with the perpetuation of socially optimal timber supplies. The origins of this myth should be well known to anyone who has completed a grammar school course in American history.

During the nineteenth century, when the United States was experiencing a period of rapid economic growth and its forests were being exposed to unfettered free enterprise, the results were devastating. Virgin forests were ruthlessly exploited for profit, cut-over lands were abandoned, streams were left choked with logging debris, and communities were abandoned in the wake of the lumber industry's "cut and run" timber harvesting activities.

No one disputes the historical record. The social costs of the lumber industry's behavior during the nineteenth century, and well into the twentieth were high. Economic dislocations are characteristic of rapid growth, and the social costs of other industries during this same period were also high.

The government gradually imposed regulations to control many of these costs, through the Pure Food and Drug Act of 1906. However, reaction to the so-called excesses of the lumber industry was somewhat different. In forestry, society has instituted regulations of private activities and, perhaps more important, has opted for public ownership of forest resources.

Today, nearly 20 percent of the nation's commercial forest lands are in public ownership. There are also millions of acres of national parks and wilderness areas that are forested, but are not included in the commercial forestry category. The latter only includes lands from which commercial timber harvest is anticipated. Parks and wilderness areas are not logged. It is important to distinguish between public ownership of forest resources to produce timber crops and public ownership to produce benefits that might reasonably be construed as public goods. Recreation and protection of the "unique" features of a natural environment are commonly classified as public goods. Whether or not this classification is appropriate is beyond scope of the present argument.[3]

The public ownership of forest resources to produce timber crops reversed the previous policy of transferring forest resources into private ownership, where their management would be subject to the dictates of the marketplace. Moreover, the management of the national forests was to be based on the sustained yield-even flow concept of timber harvest regulation. This concept is neither capitalistic nor socialistic in origin. It is, in fact, a product of feudalism.

Sustained Yield-Even Flow

Sustained yield-even flow forest management ostensibly accomplishes two worthwhile social objectives: (*a*) it provides perpetual timber crops, preventing a timber "famine"; and (*b*) it promotes community stability, precluding "cut and run" timber harvesting practices that left abandoned communities in their wake.

The Forest Service accomplishes these objectives by designating a forested area and then setting the rate of harvest equal to the rate at which timber is being grown. Once a forest is in a regulated condition, not only is the annual allowable cut equal to annual growth, but timber is also harvested at the age of "culmination of mean annual increment" to maximize average annual growth over time. This process gives rise to "maximum sustained yield." Growth can be increased by adding factor inputs to land, but it is generally assumed that these inputs are held constant over time.

If it is assumed that all acres in the forest are equally productive and that cut-over areas are reforested immediately after harvest, then total forest area divided by harvest age determines the acreage harvested each year. Moreover, the age distribution on equivalent acreages in the regulated forest will range from one-year-old trees to those that are currently ready for harvest.[4]

Many publicly owned forests are years, and in many cases decades, away from being regulated, and they contain many acres of over-mature timber. Management's task is to convert these unregulated forests to a regulated condition. In order to do this, access to the entire forest is required, and over-mature timber must be harvested. Herein lie some of the problems with the sustained yield-even flow approach.

Public foresters have generally attempted to develop access as expeditiously as possible. They oppose "progressive harvesting" or "logging off the face" of the over-mature timber inventory. If growth is to be maximized, it is important that the oldest timber stands be harvested first. Economic and environmental costs enter into management planning at this point only as ad hoc allowances. To explain why these costs are not explicitly considered, it is useful to review the origins of the approach and its implications for a market economy.

How We Acquired Sustained Yield

During the early years of this country's development, forests and timber could hardly be considered a scarce resource. Indeed, forest land had to be cleared to provide home sites, farm land, and roads. In this sense, timber had a negative value. Timber did not become scarce, except locally, for nearly 250 years, and prices remained low until after World War II.[5]

The exploitation of forests was highly predictable. The survival of producers in a market economy depends on profits. Lumber producers who were confronted with a choice of remaining in a location and growing timber to sustain a permanently located mill or of moving to areas where cheap timber supplies were available were forced by economic necessity to choose the latter. There is little evidence that the first alternative would have enhanced chances of survival. From an economic standpoint, there was a serious flaw in the conclusion that a timber "famine" would occur if the production of timber were entrusted to the market. Proponents of public ownership and management of forest lands on the basis of sustained yield-even flow failed to understand how a market economy would respond to the expected shortages.

Reasons for this belief are not hard to find. Many of those who led the crusade for public forests had received forestry training in Europe, where forests had been in the commons for centuries. Under the feudalistic systems that prevailed, private property rights (as they exist in a market economy) were largely unknown. Land resources were allocated on the basis of rights, privileges, and obligations, not by exchanges in the marketplace.

In an environment of common resource usage, there is no incentive, other than pressure that may result from social custom, for an individual to practice conservation. What one person does not get, the next person will. If a person invests his time and effort, say, to plant a tree, there is no assurance that he will be able to capture the benefits of his efforts. Common use of forest lands and timber is consistent with perpetuation of timber supplies under these circumstances as long as the rate of harvest is less than the rate of growth. When the population was relatively small, this situation prevailed, and there was no danger of depleting stands.

A useful analogy is found in the exploitation of wildlife. Because of its fugitive nature and the difficulty of making it exclusive and transferable, and therefore susceptible to being managed within the framework of the market system, wildlife has been subject to common usage throughout history. As such, wildlife populations are highly susceptible to extinction unless bag limits are imposed and enforced.[6] Because of its importance as a food source, wildlife was subject to heavy exploitation during the Middle Ages. When this began to conflict with the pleasures of the King's chase, rights to common use were extinguished, and forester-gamekeepers were sent out to enforce the rules. They also undertook population studies to determine bag limits that were consistent with perpetuating wildlife populations.

As population increased in Europe, timber harvest from common forest lands rose until it exceeded the growth rate. It became necessary to place restrictions on timber harvests similar to those that had been adopted earlier for game. These restrictions took the form of "annual allowable cuts" (bag limits on trees) and eventually evolved into the principle of sustained yield-even flow forestry.

The creation of the national forests and sustained yield-even flow management represents a failure to distinguish between an institutional framework appropriate to the exploitation of privately owned resources and one necessary for the perpetuation of common property resources. Our early foresters wrongly concluded that the timber inventory adjustment was in fact evidence that we were enroute to the depletion of all timber stands. To introduce the wrong set of institutions to "correct" a problem that was "self-correcting" within the framework of existing institutions was to lay the groundwork for many economic and environmental problems. Since the foresters' model did not incorporate costs, they had a tendency not to count them, at least not correctly.

Nothing in this argument conflicts with the possibility that social welfare may be enhanced if trees are grown for reasons that are not reflected by demands in the marketplace. Maintaining forests for erosion control, protection of water supplies, and aesthetics are all cases in point. It would be more than charitable to argue that these reasons, rather than a lack of faith in the market, were the primary reasons for creating our system of public forests.

The Allowable-Cut Effect

If a forest is managed on the basis of sustained yield-even flow, timber harvest rates are directly linked to rates of timber growth. It follows that if growth rates can be increased, then harvest rates of mature timber can also be increased. This is known as the "allowable cut effect" (ACE).

> Traditionally, [forest] managers have treated each acre independently. On each acre of land we would have to wait 40 to 60 years to harvest the results of our labors. Our [ACE] concept and treatment of returns on investment does not require such a delay. We treat our entire ownership as a production plant, each acre contributing to the whole. Under this concept the results of investments that increase the productive potential can normally be harvested immediately. The equivalent anticipated growth on one acre currently too young to harvest can be removed in merchantable harvest on another acre.[7]

Despite the intuitive appeal of this statement, it and others like it are the source of some confusing, costly, and controversial problems in public forest management.

If a forest is managed on the basis of sustained yield-even flow, then any increase in the harvest of mature timber requires an increase in the growth rate. This can be accomplished by various cultural practices such as planting, thinning, or fertilizing, or by adding acres to a forest that already supports or can be planted with growing timber stands. Public foresters frequently argue that if they can undertake the latter, they have "earned the right" to harvest more timber. If they can increase the rate of new investment (grow more timber), then they can increase the rate at which past investments (mature timber) are liquidated. This follows from the even flow constraint, and it is correct. The foresters err when they count the revenues generated by liquidating past investments (mature timber) as returns to new investments undertaken to increase growth. In effect, they are treating a point input-point output investment that will not mature for years as if it pays off as an annuity. These annuity payments are returns to a different investment.[8]

Suppose that a new investment will produce 1,000 units of wood that will be ready for harvest fifty years hence. Average annual growth generated by this investment is 20 units per year. If the forest contains mature timber stands, then, under the even flow concept, the harvest can be increased by 20 units per year for the next fifty years. The maturation period for the new investment is fifty years, but under ACE it is treated as an annuity that provides an annual return for fifty years.

The effect of ACE on estimated rates of return depends on the time period of the investment and the actual rate of return. ACE returns increase rapidly with both time and actual rate of return (see table 5), and are substantially higher. For example, a point input-point output investment with an expected rate of return of 4 percent and requiring sixty years to mature would, according to ACE, have an expected rate of return of 61 percent.

Many investments in forestry, such as site preparation, planting, and precommercial thinnings of densely stocked young timber stands, have maturation periods of sixty years or longer. With an expected rate of return of only 4 percent, these investments would not be viable today. Most public agencies are using discount rates of 10 percent to prepare their capital budgets. But with ACE, the 4 percent, sixty-year investment is made to appear highly lucrative.

Most public agencies that manage commercial forest lands incorporate ACE calculations into their management plans, although they have recently tended to report them as benefit-cost ratios rather than

TABLE 5. ACE Rates of Return for Various Actual Rates of Return and Time Periods

Time Period (in years)	Point Input-Point Output Actual Rate of Return			
	4	6	8	10
20	11	17	25	37
40	12	28	61	136
60	18	61	203	672
80	31	152	752	3726

Note: The actual rate of return from a simple point input-point output investment can be determined from the relationship:

$$C_o = PQ_e{}^{-rT}, \tag{1}$$

where

C_o = investment cost
P = price per unit of output
Q = quantity of output at time T
T = maturation period of C_o
r = expected rate of return to C_o.

As a result of making C_o, average annual growth on the forest will be increased by Q/T units per year for the next T years. Where ACE is used, harvest rates of currently mature timber can, therefore, be increased by Q/T units per year for the next T years. This gives rise to an annuity of the form:

$$C_o = \frac{PQ'}{T} \left(\frac{1_{-e}{}^{-iT}}{i} \right), \tag{2}$$

where

Q' = Q units of currently mature timber
i = rate of return on annuity

If T is large, which is reasonable to assume for most forestry investment, then equation [2] is approximately

$$C_o = \frac{PQ}{iT}. \tag{3}$$

ACE rates of return are derived by setting equation [1] equal to equation [3] and then solving for i:

$$PQ_e{}^{-rT} = \frac{PQ'}{iT}.$$

Since

$$Q = Q'$$

$$i = \frac{e^{rT}}{T}.$$

Table 5 contains values of i (ACE rates of return) for various values of r (actual rates of return to C_o) and T.

as expected rates of return. The results are the same, however. Benefit-cost ratios are exorbitantly high: many are over 10:1, and some in excess of 50:1 have been reported.[9]

Economists, financial accountants, and others who are exposed to these calculations for the first time find them unbelievable. Nevertheless, they are being used by forestry officials to justify budget increases to grow more timber and, therefore, to harvest more timber.[10] In addition, industry lobbyists commonly use ACE arguments to get Forest Service budgets increased. Many in the industry are facing extremely tight timber supplies, and they are likely to be faced with mill closures. Given a choice of supporting nonsense that may get them more timber now or working for legislative or administrative reform that may take years, their actions are not surprising.

Public foresters have been almost impervious to criticism of ACE calculations. They argue that if the even flow constraint is there by law, and if, therefore, the only way that timber harvests can be increased is by growing more timber, then receipts from increased sales cannot be had unless investments are made to increase growth.[11] These receipts are, then, appropriately used to estimate the rate of return on the new investment.

While it is correct that a strictly enforced even flow constraint requires new investments before timber harvests can be increased, it does not follow that receipts from this increase should be used to estimate the rates of return. Rates of return can and should be estimated correctly so that the general public can know what the even flow constraint is costing. Public agencies and their supporters are understandably not anxious to have this happen.

Cross-Subsidizing

It would be difficult to estimate the magnitude of the waste that might be caused by ACE, but it is not likely that it is as great as that which results from using existing public timber wealth to cross-subsidize uneconomical forest management activites. Much of this takes the form of tying public timber sales to the construction of roads.

As noted earlier, an objective of sustained yield-even flow is to maximize growth. In order to do this, decadent stands with negligible rates of growth must be harvested, and new stands must be established. Many of these older stands are at higher elevations and in the more remote areas of the national forests. In order to get roads into these areas, Forest Service officials can specify in timber sales contracts that the removal of timber is made contingent on the construction of roads. By this process, the Forest Service is able to effectively circumvent the budgeting process and to spend existing timber wealth to implement uneconomical sustained yield-even flow activities.

Environmental impacts of these activities are also substantial, because forest access roads are known to be an important source of erosion, and hence stream pollution. By its policy of making entire forest areas accessible, the Forest Service is creating greater environmental impacts in terms of erosion and degradation of water quality than would be the case if it pursued a policy of maximizing present net worth of the resources it manages. If the latter objective were to guide its activities, then the Forest Service would maintain a less extensive road system, and both the costs of road construction and the adverse environmental impacts would be reduced.

An additional factor is the visual effect of timber harvesting activities. With an extensive road system, timber harvests tend to be scattered in smaller areas around the forest. If the road system were less extensive, harvests would tend to be more concentrated. It is a matter of personal preference whether one finds concentration of timber harvesting activities more aesthetically pleasing. It is a matter of simple economics, however, that maximum dispersion of harvesting activities is more expensive. More roads are required, and they are costly to construct and to maintain.

This raises the question of whether the use of timber wealth to finance extensive road systems might not be producing enough reductions in visual disamenities to be worthwhile. The question is difficult to answer, largely because it has not been thoroughly investigated. The Forest Service has initiated a practice called "aesthetic logging" (a contradiction in terms to many), by which the boundaries of logging sites are located according to topography, ridges, valleys, rock slides, etc. The purpose is to give the logged area a "natural" appearance. Previously, areas to be logged were generally located by section lines. Where this grid survey system is used, hillsides often take on the appearance of patchwork quilts.

Aesthetic logging may reduce visual disamenities, but it is costly. Where it is practiced, logging costs are higher; hence, prices that public agencies receive for the timber they sell must fall. Does the public receive additional environmental benefits commensurate with the additional costs? Unfortunately, the only answers we have to this question are statements by Forest Service officials that the general public wants these benefits; therefore, they have a responsibility to provide them. To be sure, the Forest Service and other public agencies are under pressure to provide these benefits, but it is not obvious that this pressure comes from a representative cross-section of the general public. Indeed, there are good reasons for believing that public timber wealth is being used to try to silence critics of timber harvesting activities. Public hearings on public timber management practices and the Forest Service's public involvement programs tend to attract the same people and the same groups. It would be difficult to claim that they represent the general public.

About the best that can be done under present circumstances is to ask for accountability. What is being done, and what is it costing? Washington State's Department of Natural Resources (DNR), in its annual report for 1974, reported that environmental restrictions on logging operations had increased logging costs by $14 per thousand board feet (Mbf). This was about 10 percent of the price that the Washington State DNR was receiving for the timber it sold that year. The means by which this estimate was made were not discussed, which leads one to believe that it was based on informed opinion. Nevertheless, it did represent an attempt to inform the general public that environmental restrictions on logging operations are costly. The DNR, it should be noted, has a legislative mandate to maximize net income to the trusts whose timber it manages. The Forest Service has no such mandate. Its mandate is to produce multiple use benefits on a sustained yield basis.[12]

The ambiguity of the Forest Service's mandate has permitted this agency to escape accountability based on cost-effectiveness analyses of the wealth it is managing. It is almost sufficient for Forest Service officials to assert that something is in the public interest for them to justify using public timber wealth to help finance it. With more attention being focused on the Forest Service and its effects on the forest products industry, prices, income, and employment, this situation will quite likely change. Taxpayers will also be directly affected. Without a healthy forest products industry, public timber is worth less; hence, timber management policies that create an anemic industry are another means of squandering public timber wealth. There are good reasons for believing that problems of these kinds will continue to occur.

How to Create an Anemic Industry

Over the years, a disparity has arisen between public ownership of commercial forest land and commercial timber inventories. While only 20 percent of the nation's commercial forest land is in public ownership, 50 percent of the nation's softwood sawtimber industry is publicly owned. In the West, the percentage is much higher—over 70 percent in Idaho, Oregon, and Washington, for example. The softwood sawtimber inventory is the primary raw materials base for the lumber, plywood, and pulp industries.

The manner in which these publicly owned timber inventories are managed will obviously have a profound effect on the economic viability and the structure of the forest products industry for decades to come. Without substantial changes in the sustained yield-even flow policy and the way public timber is sold, there are good reasons for believing that the industry and those who depend on it will be adversely affected.

The forest products industry developed around privately owned timberstands. This was to be expected, given that most of these resources were over-mature. It was in the owners' interests that this timber be liquidated and the lands converted to higher value uses or planted with new timber stands.

As a result, capacity in the timber processing industries has been based on private timber supplies, plus expectations of what the public sector would supply following the liquidation of privately owned virgin timber and prior to the maturing of second-growth timber crops, which were established on cut-over lands. Until privately owned virgin timber stands were liquidated, public agencies had very little market for their timber.

Now that privately owned, virgin timber is nearly depleted, industry demands for public timber have risen rapidly. At the same time, demands are being made on the public forests for the production of more recreational benefits and environmental amenities. Trade-offs between the production of timber and recreational and environmental benefits have resulted in incorrect estimates of public timber supplies. One result has been an excess capacity in the processing industry relative to total timber supplies, public and private. Recent mill closures in the West provide examples of the adjustments industry has had to make because of the environmental movement.

Environmentalists have seized this shift in demand to argue that the industry has cut all its own timber and now it wants to cut the public's. The expected sequence of harvesting private and then public old growth followed by private and then public second growth is, as would be expected, ignored in their argument. This approach has helped environmentalists promote their cause, although it has been, and apparently will continue to be, a considerable hardship on the industry.

The way in which public timber is marketed is another aspect of the sustained yield-even flow policy that has been detrimental to industry. Even flow, it should be noted, is not a harvesting policy; it is a timber sales policy. A relatively constant amount of timber is sold each year, mostly by auction bidding; but the amount harvested each year may be quite different.

The responsibility for promoting stability is shifted from public agencies to timber operators as soon as public timber is sold. Operators have the choice of timing their timber harvests within the limits of their timber sales contracts. In periods of rising timber prices, timber operators will speculate in their bidding. Then, when they acquire the sale, they hold it until prices rise, at which time it is harvested and sold. If prices fail to rise as much as anticipated, they may hold it longer to avoid being left with the difference between what they bid and the resale price. In any event, the system is highly uncertain, a situation that is compounded by the fact that most public timber sales are relatively small and short term.

The policy of limiting sales is dictated largely by political considerations: small operations must be given an opportunity to compete with larger ones. The overall problem with this policy is that it discourages large investments, which are necessary to exploit economics of scale in timber processing. No one seems willing to make investments in mills that would maximize returns to public timber.

Present sales policies are already counterproductive to the stability objective they were designed to achieve, as evidenced by fluctuations in unemployment rates in communities that depend on public timber. The situation is likely to get worse as industry capacity shifts toward mills that are most efficient at playing the speculative game.

Much of the new investment in the timber processing industry is being made in small log mills that are designed to process second-growth timber from the private sector. These mills cannot process the large, virgin timber that characterizes public timber inventories. By discouraging investments in the mills that can process public timber efficiently, public agencies are causing demand changes that will adversely affect prices, and hence the timber wealth that they manage.

Conclusion

The sustained yield-even flow concept has managed to survive because of public misunderstanding, the political influence of the forestry profession that developed it, and the fact that at the time the system of public forests was established, timber was so plentiful that opposition to taking some of it off the market was not as great as it might otherwise have been. Sustained yield-even flow had years to become institutionalized before its compatibility with the market system was really tested. Now that we are getting results from these tests, the concept is increasingly being found to be an institutional failure.

The major problem with sustained yield-even flow and its corollary ACE is that they waste public timber wealth. In addition, maintaining institutional arrangements based on this approach has been very expensive for taxpayers. The Forest Service especially incurs deficits each year that currently average about $500 million and are expected to rise in the near future to more than $1 billion per year.

While the Forest Service and its supporters are quick to note that the national forests produce many nonmarketed benefits, this does not preclude the fact that, if they were being managed by economically rational criteria, society could have more wood and no less nonmarketed benefits, more nonmarketed benefits and no less wood, or more of both.[13] More timber could be cut, but it would be cut in different locations so that conflicts between timber harvesting and environmental amenities would be reduced.

Notes

1. The $42 billion estimate of the value of the national forests was developed by Marion Clawson, "The National Forests," *Science* 191 (1976):762–67.

2. The Forest Service is not the only agency that manages publicly owned forests and also subscribes to economic nonsense of the kind described. Other federal agencies, such as the Bureau of Land Management and the Bureau of Indian Affairs, both in the Department of Interior, and several state agencies do, too. Criticism is directed toward the Forest Service because this agency is the largest, and it has tended to be the style setter for the rest.

3. For some alternative views on this issue, see John A. Baden and Richard L. Stroup, "Private Rights, Public Choices, and the Management of National Forests," *Western Wildlands*, Autumn 1975, pp. 5–13; and Edwin G. Dolan, "Why Not Sell the National Parks?" *National Review* 23 (1971):362–65.

4. For a more complete description of the sustained yield-even flow model, see a forest management textbook, such as Kenneth P. Davis, *Forest Management* (New York: McGraw-Hill, 1966), or Thomas R. Waggener, "Some Economic Implications of Sustained Yield as a Forest Regulation Model" (Institute of Forest Production, University of Washington, Seattle, 1969); or LeRoy C. Hennes, Michael J. Irwing, and Daniel I. Navon, "Forest Control and Regulation: A Comparison of Traditional Methods and Alternatives," Research note PSW–231, U.S. Department of Agriculture, Forest Ser-

vice, 1971. Public foresters frequently defend their approach to forest management by observing that their critics have not considered it in proper detail. While this may be a useful debating tactic, it makes scientific discourse more difficult by denying access to Occam's razor.

5. Douglas fir stumpage (standing timber) prices were $2.20 per Mbf in 1910. They fluctuated very little until 1940 when they were $2.30 per Mbf. By 1950, they had risen to $16.40 per Mbf. The flow of resources into growing timber increased rapidly after 1950. For historical stumpage price data, see U.S. Department of Agriculture, Forest Service, "Price Trends for Forest Products" (Washington, D.C., 1956).

6. If wildlife resources range within the limits of a given sovereignty, then it is misleading to say they are common property resources. They belong to the state and in this sense they are exclusive and transferable. What the state chooses to do with them is a separate question. They may be managed as common property resources because that is less costly than enforcing exclusive and transferable rights so that they can be privately owned and managed within the framework of the market system. On the other hand, for reasons of political expediency, the state may continue to regulate their use as common property resources even though it would be less costly to enforce exclusive and transferable rights to them, place them in private ownership, and permit them to be managed by the market. The situation with fish and wildlife is not unlike that in public forestry. Even though the existing institutional arrangement is uneconomical and inequitable to maintain, it may be difficult to change because of vested interests. Public reaction to being exploited by these interests may be necessary before desirable change can be implemented.

7. U.S. Department of Natural Resources, "Sustainable Harvest Analysis," Harvest Regulation Series, Report no. 3, 1970, p. 3.

8. Quite rightly, public as well as private foresters have a problem of determining the appropriate unit to consider in evaluating their management alternatives. If it is inappropriate to look at the overall forest, why should they stop at individual acres? Perhaps they should look at individual trees. Observations such as these are used to justify continuing present policies. The appropriate size unit to consider is an economic question. The costs of increased cross-subsidization within units if larger units are evaluated must be weighed against the increased costs of using smaller units in the analysis. Under some conditions, this might mean evaluating individual trees. Under other conditions, it could mean evaluating forty-acre or larger tracts.

9. U.S. Department of Natural Resources, "Sustainable Harvest Analysis," Harvest Regulation Series, Report no. 6, 1975.

10. See, for example, testimony by Chief McGuire of the Forest Service before the House Appropriations Subcommittee, 1976.

11. Public foresters have helped to write much of the legislation by which public lands are managed. Much of this legislation was written at a time when public forests were not nearly as important as they are today; hence, their

ideas were easier to get written into law. Today, they frequently take refuge in these laws to avoid debating the merits of their ideas. Their defense of ACE, even though it is improper accounting, is an excellent case in point.

12. Multiple Use-Sustained Yield Act of 1964.

13. In simple economic terms, sustained yield-even flow criteria keep society inside its production-possibilities curve for wood and the environmental benefits with which wood production conflicts.

Compounding Clear-cuts: The Social Failures of Public Timber Management in the Rockies

by William F. Hyde

While clear-cutting has been the most visible interference with amenity valued forest resources, a potentially greater problem exists. Historically rational public timber management policies are no longer economically efficient. They result in large misallocations of public funds and both timber and amenity valued forest resources.

Our public forests include some highly productive timber lands. They also include some timber land that is less productive—for both biological and economic reasons—even though this land may have a standing timber inventory. For either class of land, there are large social costs incurred when managers ignore economic efficiency criteria.

Less productive timber land includes the steep slopes, shallow and unstable soils, high elevations, and exposed vistas, all of which are associated with (*a*) environmentally risky timber management and (*b*) high values for conflicting nontimber uses, such as wilderness recreation. Thus, expansion of timber management activities to less productive land can be reflected in environmental destruction and foregone recreational opportunity.

Consider three timber management options: (*a*) continuing current Forest Service practices, (*b*) allowing a free timber market to operate, and (*c*) managing for economic efficiency. Economic efficiency describes the social optimum and differs from the free market option in that it affects values that fail to exchange in the market. It is our contention that free timber market results more closely resemble

This article was written while the author was at Resources for the Future, Washington, D.C.

the socially efficient optimum than do the results of current practice. Specifically, *disregard for the free timber market leads to a misallocation of public funds in favor of expanding timber management activities to less productive land. This misallocation is reflected in foregone nonmarket environmental and recreational values, which characterize Forest Service management of over five million acres in Colorado.* These Colorado forests are generally conceded to be of poorer quality for timber but outstanding for hiking, camping, hunting, fishing, skiing, and other amenity values.

We preface our examination of this hypothesis with the historical rationale for timber production in Colorado and the justification for Forest Service intervention in the timber market. Subsequently, we examine a simplified version of the agency's approach to timber harvest and investment decisions and then introduce our case study. The area for the case study is the San Juan National Forest, a 1.85 million acre unit in southwestern Colorado on which annual harvests have averaged 70 million board feet (MMbf). The Wilderness Society and the Colorado Open Space Council have appealed an initial Forest Service decision to expand harvest to 125 MMbf by including harvests from four major and several lesser roadless areas within the forest.

We examine the merits of the Forest Service decision and the environmentalists' appeal from the perspective of (*a*) free market timber impacts, contrasting these impacts with those that obtain under (*b*) current Forest Service management. Not all other forest uses conflict with timber harvesting, but where *negative* timber values exist, reliance on the timber market would reduce production and expand other, conflicting uses to more accurately reflect the social optimum. Since land is efficiently allocated when put to its highest valued use, conflicting forest uses must exceed *positive* timber values in order to obtain priority over timber management. Following our discussion of the San Juan National Forest, we comment on broader implications of our analysis and on recent changes in Forest Service policy.

Historical Background

Two factors combined to make logging economically viable in nineteenth-century Colorado: a plentiful inventory of mature timber and ready markets for mine props and railroad ties. Early logging was not accompanied by reforestation, however, and natural regeneration is slow in the arid Colorado climate. The result is a less plentiful inventory today, with some former timberland still in brush and grass

100 years after it was logged. Prices originally justified harvest of the natural inventory, but they did not justify the investment necessary for sustained production. Meanwhile, the mine and railroad markets have disappeared. To be sure, they have been replaced with an expanded local housing market, but this corresponds to a larger population also known for its recreational forest demand. Thus, a timber market that remains important, although less significant for Colorado's economy than it once was, is now competing with increasingly strong demands for the state's limited forest resources.

Logging and the lack of regeneration eventually led to a national perception of future timber shortage, which resulted in the creation of the national forest system. Thus, *perceived market failure to respond to expected future values was the original justification for the Forest Service.* Moreover, perceived market failure also explains statutory and administrative policies requiring (*a*) regeneration subsequent to harvests and (*b*) perpetually sustainable harvest flows, the two primary components of the Forest Service approach to timber harvest and investment decisions.

With the knowledge of neoclassical economics, however, the case for public intervention in the timber market is no longer certain. We now understand that prices increase in response to expected shortages and that investment is a response to price expectation. That is, stumpage prices (the prices paid for standing trees) adjust so that long-term timber shortages will not occur in a free market.

The current justification for public intervention is that, *when forests are managed in accordance with a free timber market, important conflicting nonmarket values, such as wilderness recreation, are overlooked.* This may be a reasonable argument. If the Forest Service acts in accordance with it, we expect the expanded provision for nonmarket values to lead to lower harvest levels and less forest land use for timber than would occur under a free timber market (but no market for amenity valued forest resources).

Forest Service Harvest and Investment Criteria

There are three independent elements of Forest Service harvest and investment analysis: (*a*) the timber land base decision, (*b*) the timber maturity decision, and (*c*) the harvest and investment decision itself.[1] The timber land base is determined as a residual and has no particular economic justification. It is what remains after nonproductive forest land, the surface area of water, and those small amounts of *productive* land reserved for special uses (such as administrative sites and wilderness, scenic, and geologic areas) are subtracted from the total land area managed by the Forest Service. The remainder, called commercial forest land, is approximately that which can produce twenty cubic feet per acre per year of naturally grown wood fiber. There is considerable evidence that some of this residual is incapable of yielding an economic rent in timber production[2] and, therefore, is inefficiently allocated to timber, even if conflicting nonmarket values are disregarded. If this is correct, there is an *overinvestment in land for timber production* and, correspondingly, an increase in environmental risk resulting from overproduction of timber along with a loss in production of conflicting nontimber forest resources.

Maturity is the age at which a timber stand becomes eligible for harvest. Economic or financial maturity is a function of expected stumpage receipts, management costs, and the alternative use of funds or the discount rate. It is attained when the net revenues from harvests exceed the discounted expected net revenues from delaying harvests one year. The Forest Service, however, sets maturity at the age when average annual physical product (growth) is maximized. The two definitions are coincidental only on economically productive timber land when the discount rate is zero.[3] Therefore, the *Forest Service definition of maturity implies an excessive investment in capital, or older and larger timber stands.*

Once the Forest Service makes independent determinations of the timber land base and the age of timber at maturity, it makes its harvest and investment decisions according to the excess inventories of mature timber. Annual harvests are set to equal average annual growth, with excess mature timber rationed over the long run so that an even annual flow of timber *volume* obtains from year to year. By this rule, annual investment on one parcel of the timber land base implies increased growth on that parcel and, therefore, allows increased harvests from the forest as a whole. Since the harvest increase can originate only from the excess inventory of mature timber, this rule has harvests in one location justifying investments in another. In some cases, wise harvest decisions may subsidize unwise investment decisions. This Forest Service practice contrasts with the economic rule that would treat management on different parcels as *independent* decisions, requiring each to be justified by its own expected *financial* returns in excess of costs.

Prices first enter the Forest Service analysis after the harvest decision has been made. The potential harvestable volume is appraised, and if its appraised price exceeds an arbitrary minimum (base) price, the timber is auctioned to the highest bidder. The economic rationale for this procedure is that almost any positive appraised price exceeds short-run costs, which are only the costs of timber sale administration. A wise producer, however, might make the sale and then use the sale price, modified for future expectation, to guide his reinvestment decisions. If the expected revenues implied by this price are insufficient to cover long-run costs, i.e., all the costs of timber management over the full timber growing period, then the wise producer cuts his losses after this first sale and exits from the timber producing industry, allocating his forest land to another use.

The market is ordinarily a stern teacher, restricting repetition of such errors by driving those who make them out of business. The Forest Service, however, is protected by an excess inventory of mature timber. Receipts from additional sales appear to balance financially unrewarding decisions elsewhere. Moreover, the Forest Service ignores prices in most of its decision making, and where it uses them, it fails to use them as learning signals. As a result, our expectation is that the Forest Service overinvests in timber land and in silvicultural (timber growing) practices (to the detriment of environmental values) on some locations and delays timber harvests beyond the economically justifiable time in others. This conflicts with free market efficiency. Moreover, it opposes the argument for public intervention in order to provide for nonmarket values. Apparently, allowing a free timber market to operate would provide for both timber and some environmental values at a level more closely approximating their social optimum. In our examination of the San Juan National Forest, we confirm this finding.

Southwestern Colorado is a country of mountains, mesas, canyons, and rivers, spotted with abandoned mines and prehistoric cliff dwellings. Tourism and ranching support the local economy, although there is also some logging for sawmills in two local towns (combined population 12,000). The isolation of the area has minimized timber production and restricted recreational development.

Over the years, the local mills have depended on virgin stands of mostly ponderosa pine from the San Juan National Forest. As these stands have disappeared, harvesting has progressed up the mountainsides to include greater proportions of lower valued Engelmann spruce, with some aspen and Douglas fir. In recent years, an average of 70 MMbf per year (55–70 percent spruce) have been harvested, although current sales average less than 50 MMbf per year. (The difference between harvests and sales is the result of large sales in 1963 that have not made their way to the mill until more recently.)[4]

To maintain an annual harvest level of 70 MMbf, the Forest Service plans to harvest four large and fifteen smaller roadless areas in the next five years.[5] Herein lies the major environmentalist objection to the San Juan National Forest timber management plan. These roadless areas are generally classified by the Forest Service as "marginal-access," which means that the logging roads necessary for their harvest cannot be justified by the standing timber values. Disregarding the apparent efficiency costs of maintaining this harvest level and the fact that 30 percent of the timber offered for sale in the past five years has not even been bid upon,[6] the Forest Service plans to expand annual harvests to 125 MMbf. The entire expansion will originate from land classified as either marginal-access or marginal-logging. Marginal-logging lands are those with critical soil or topographic conditions that require expensive balloon, helicopter, or cable logging systems in order to prevent environmental degradation. All told, the expanded harvest level requires 1,130 miles of new roads, including 570 miles on marginal-access lands, and another 1,285 existing miles that will have to be upgraded for logging.[7]

Approach

In the case of the San Juan National Forest, timber values are somewhat difficult to assess. The Forest Service does not maintain a functional budgeting system comparing all costs and receipts associated with timber, let alone a project account for each timberstand or independent land management unit. In the absence of such systems, we must develop our own estimates of local timber production costs, prices, growth, and yield. From these, we can estimate the harvest levels that might obtain in a free timber market and compare them with the San Juan National Forest timber management proposal. We use high estimates of prices and yields and low estimates of costs to present an optimistic view of timber management opportunities. Moreover, such a view introduces a conservative bias into estimated losses resulting from excessive timber production where timber conflicts with other forest resource uses. It prevents us from overstating any case against timber management.

At this point, it is appropriate to more carefully examine the relationships between timber and other forest resource uses. Some resource uses, such as wilderness recreation, may conflict with timber production. Others, such as water and wildlife resources, may be compatible with timber production. Where compatible uses exist, it is inappropriate to charge the common costs of their production, such as administration and protection costs, entirely to timber. Rather, we follow the theoretically correct rule of charging to each compatible use only its separable costs, or

$$B^T - C_s^T \geq 0, B^{NT} - C_s^{NT} \geq 0,$$

where B^T, B^{NT} = timber and compatible nontimber benefits, and C^T, C^{NT} = timber and compatible nontimber costs. Timber production is positively valued in this case if expected timber receipts exceed separable timber costs, and if

$$\sum_i (B^i - C_s^i) - C_c \geq 0,$$

where C_c = common, or inherently inseparable, costs. The sum of excess separable costs for all compatible uses, including timber, exceeds the common costs.[8]

The Multiple Use-Sustained Yield Act of 1960 and the National Forest Management Act of 1976 require that timber harvests be sustainable for the long run, and the Knutson-Vandenberg Act of 1930 forbids timber mining, which means harvesting virgin timber and then benignly neglecting the land.[9] Therefore, a positive net perpetual timber value is a necessary condition for efficient public timber management. Perpetual timber value is the present value of successive growing cycles (rotations) discounted to take into account that expected profits from future harvests are worth less because profits that are realized at present can be invested to earn additional interest income over time. Timber stands having a positive net perpetual timber value pass the *long-run efficiency test*.

For previously unmanaged timber stands, initial period costs and receipts, comprised primarily of the initial capital outlay for road building plus receipts from the sale of the large (virgin timber) harvests, may sharply affect the direction of the perpetual timber value. The benefits of road construction may extend to subsequent harvests, but the discounted value of these benefits is small, even for harvests occurring as soon as fifty years, and we can operationally charge road construction costs and benefits entirely to the initial harvest. Therefore, for previously unmanaged timber stands, we assume a rule of separating initial value from perpetual value. A positive net value for initial period activities becomes a second necessary condition. We call this the *short-run efficiency test*. Together with the knowledge that net timber values exceed conflicting forest resource use values, long-run and short-run efficiency are sufficient conditions for timber management on public lands.

Analysis

In this section, we estimate free market timber management values and assess the costs of current Forest Service management on the San Juan. First, we examine initial period market values for previously unmanaged timber stands. Second, we examine perpetual timber values, first for the case where timber is the primary output, then where timber is produced jointly with other compatible forest resource uses. Finally, we contrast these perpetual timber values with those that obtain under current Forest Service management procedures. All revenues and costs are in 1976 dollars, reflecting the publication date of the San Juan National Forest Timber Management Plan.

Initial Period Values

The Timber Management Plan calls for the introduction of timber harvesting on 169 thousand acres that are marginal due to access conditions and 249 thousand acres that are marginal due to logging conditions.

On land that is marginal due to access, the Timber Management Plan calls for annual harvests of 44 MMbf on 11,200 acres, or 4 Mbf per acre. Initial period costs of obtaining this harvest include (*a*) the capital costs of road building, (*b*) the cost of timber sale administration, and (*c*) the regeneration costs. Road building costs are the largest single item. By definition, sites that are marginal due to access are unroaded. New roads on these sites cost in excess of $30 per Mbf of timber removed from the site.[10]

The Forest Service absorbs the cost of timber sale appraisal. Similarly, it absorbs the cost of sale administration, but we were able to obtain an estimate of $8.70 per Mbf for the latter.[11] We shall find that even this cost would be a severe handicap to free market management for timber on the San Juan National Forest.

The Knutson-Vandenberg Act requires that all land, once harvested, be regenerated. In fact, it requires that funds for necessary replanting be set aside from the receipts of the previous harvest. The timber management staff of the San Juan National Forest maintains that, historically, natural regeneration has been largely successful.[12] Furthermore, current harvest methods are chosen to obtain 100 percent natural regeneration. Nevertheless, approximately one-eighth of the acres harvested have required artificial regeneration at a cost of $367 per acre. If even half this many acres require artificial regeneration under current harvest methods, the average cost per acre is $23, or $6.75 per Mbf ($23 per acre for 4 Mbf per acre) harvested. We will overlook the cost of occasional fencing ($2,000 per mile) required to protect young seedlings from animal damage.[13]

In conclusion, minimum initial period costs per Mbf are:

Road building	$30.00
Sale administration	8.70
Regeneration	6.75
Total	$45.45

Receipts must equal or exceed this value in order to meet the first condition for timber management. Therefore, the market stumpage price must equal or exceed $45.45 per Mbf.

The average sale price for 1976, however, was only $2.65 per Mbf. The highest average annual sale price since 1970 was only $23.15, and only twice has it exceeded $7.00 (although extreme values for a single sale vary from $1.00 to $97.00). If we consider that marginal-access timber land is generally further from the sawmill than the average timber sale and that the logger or mill owner must absorb the additional transportation costs, then the maximum price bid by loggers for marginal-access timber sales is probably lower than the average sale price. Apparently, stumpage prices rarely exceed $45.45 per Mbf, and land that is marginal due to access rarely meets the short-run efficiency test for timber management. The Wilderness Society and the Colorado Open Space Council appeals to restrict logging expansion to these 169,000 roadless acres has considerable merit on grounds of market timber values alone.

The efficiency of harvesting lands that are marginal due to logging conditions is more difficult to assess. An uncertain number of roads must be built and additional mileage must be upgraded in preparation for logging. The timber management plan calls for 560 miles of the former and 1,285 miles of the latter, but we do not know how either is allocated between the marginal-logging and the currently managed components of the forest. Even if there were no initial capital outlay for roads, however, at the rate of 4.7 Mbf per acre,[14] sale administration and regeneration costs exceed $13.80 per

Mbf. Even this level of costs is greater than the average annual sale price for timber for all but one year since 1970. Moreover, for timber sales on marginal-logging lands, someone—loggers, millowners, or the Forest Service—must absorb not only additional transportation costs, but also the abnormal costs of balloon, helicopter, or long cable logging systems. Apparently, stumpage prices rarely exceed the short-run costs of harvesting land classified as marginal-logging on the San Juan National Forest.

Perpetual Timber Values

The marginal timber lands that the Forest Service proposes to introduce to long-term management fail the initial market condition. It remains for us to examine whether perpetual, or sustained yield, timber management is an appropriate use of Forest Service land and funds, even on those acres of the San Juan National Forest currently managed for timber. In general, these lands are located at lower elevations and on gentler slopes than the marginal lands. They are also stocked with a larger proportion of ponderosa pine, a higher valued species than the Engelmann spruce that grows at higher elevations. Our expectation is, therefore, that this land is more likely than the marginal lands to satisfy the long-run market efficiency condition.

For the perpetual timber management case, we find the maximum present net value obtainable from growing timber on a given unit of land. Accordingly, we sum costs and revenues due to timber growing as they occur and discount. The standard mathematical expression, known as the Faustmann equation, is

$$V = pQ(T)e^{-rT} - C(t) \ 1 + e^{-rT} + (e^{-rT})^2 + \ldots,$$

where V = present net value
 p = stumpage price
 Q = harvest volume, a cumulative function of time or timber stand age T
 r = discount rate
 C = management costs
 T = timber production period (rotation) or age at harvest.
Periodic costs must be discounted to the beginning of the rotation and summed. The infinite series represents the present value of subsequent timber crops. It is the mathematical equivalent of $(1 - e^{-rT})^{-1}$. Therefore, the Faustmann equation is usually expressed as

$$V = pQ(T)e^{-rT} - C(t) \ (1 - e^{-rT})^{-1}.$$

We use this latter form and solve for the minimum price at which the present net value is positive; that is, the lowest price at which timber management can be justified on its own merits. Our results depend on the chosen rotation age.

Stumpage prices have shown an upward secular trend,[15] but so have competing nonmarket amenity values.[16] The Faustmann equation is correct so long as these secular trends continue at their approximately offsetting rates in the neighborhood of 2 percent per year.

Timber stand volume varies with the quality of the land and the stocking of the stand. One measure of land quality is site index. The preponderance of commercial timber land on the San Juan National Forest is between ponderosa pine site index 40 and site index 79. Considering site index 70 to be a generous estimate of average site quality, we can refer to normal yield tables for ponderosa pine to learn the age and harvest volume relationship.[17] Normal yield tables, however, refer to fully stocked stands. The average stand on the San Juan is less than 60 percent stocked.[18] Yields for 60 percent stocked stands of ponderosa pine site index 70 follow the schedule, where volume is measured in board feet (Scribner scale):

Age	50	60	70	80	90	100	110
Volume	420	1,320	2,580	4,200	6,000	7,860	9,720

The discount rate is defined by the best alternative use of public funds. The Office of Management and Budget (OMB) requires a rate of 10 percent,[19] but allows the Water Resources Council to use a temporary, negotiated rate of approximately 7 percent for natural resource projects. The Forest Service would prefer to use the latter rate. The final choice rests with the OMB, which approves Forest Service budgets.

The same sale administration ($8.70 per Mbf) and regeneration ($23 per acre) costs exist for marginal lands. In accordance with the Knutson-Vandenberg Act, regeneration costs are charged to the previous harvest at the time of harvest. The same uncertainty that existed for the marginal-logging land classification exists for road costs. That is, the timber management plan indicates a large number of both new and upgraded roads, but it fails to show the division of these between marginal-logging and currently managed lands. As a result, we know road costs are substantial, but we cannot confidently quantify them. Reinforcing the conservative nature of our timber production cost estimates, we ignore these road costs.

The only new costs are the annual costs of general administration and forest protection. These costs include such items as upkeep on the district ranger's office, fire and insect suppression, and road maintenance. Forest Service accounting practices make it difficult to establish the magnitude of these costs, but, if experience in the Pacific Northwest is representative, annual costs exceed one dollar per acre.[20] To repeat, these must be individually discounted and summed. Over the 50-to-150-year duration of a timber rotation, this sum becomes significant.

Annual administrative and protection costs represent common inputs to all uses of the land. They can be charged entirely to timber production where timber is the primary output and all other resource uses are simply by-products. A reading of the San Juan Timber Management Plan suggests that timber is the primary output on the lands currently managed for timber. Nevertheless, where timber is one of several important outputs, timber receipts need only cover the separable costs of timber production. In this case, timber management is justified if the sum of separable benefits from other forest resource uses exceeds their separable costs by more than the costs of common inputs. If excess separable benefits are insufficient to cover the common costs, the appropriate Forest Service role is that of protecting idle land. In this case, the Forest Service should allow no exploitation and no irreversible use of the land, protecting it until relative values have changed so that some resource use is efficient. Meanwhile, *de facto* wilderness management is appropriate.

Using the preceding data and the separable-common cost allocation, we can determine the minimum stumpage price at which timber production becomes efficient. We expect this price will vary with the discount rate and according to the share of common administrative and protection costs that timber management has to bear. The results of our analysis are shown in table 6. They bear out our original contentions.

Where timber management is the primary use of the land, the stumpage price must be sufficient to cover all the costs of land man-

TABLE 6. Economic Criteria of Perpetual Timber Management

Discount Rate of Minimum Stumpage Price (per Mbf)	Timber Primary Output (in dollars)	Multiple Outputs (in dollars)
7%	80	26
10%	260	26

agement. It must exceed $80 per Mbf if the discount rate is 7 percent and $260 if the rate is 10 percent. In both cases, the optimal harvest timing occurs when the timber stand is sixty years old and contains 1.32 Mbf of timber. That is, $T = 60$ maximizes the Faustmann equation. When there are a multiple of forest resource uses and the combined value of the compatible nontimber uses exceeds their own separable costs plus *all* the common annual costs of administration and protection, then the stumpage price need only exceed $26 per Mbf, which is essentially the cost of harvesting. In a more usual multiple output case, we would expect timber management to share the burden of common costs.

Referring once again to average annual sale prices that have really exceeded $7.00 per Mbf and have never exceeded $23.15 since 1970, apparently *only rare high-valued timber sales are justified. Timber management fails both short- and long-run efficiency tests on both marginal and currently managed San Juan National Forest land.*

Current Management

In the preceding analysis, we assume that economic criteria guide the harvest timing decision. The Forest Service, however, prefers physical criteria, harvesting when average annual physical product is maximized—approximately 100 years for ponderosa pine. We might contrast the costs and minimum offsetting stumpage price using physical criteria, with the costs and price relevant to market criteria.

Previously, we assumed stands were fully harvested (clear-cut) at one time. The Forest Service (for ponderosa pine) prefers to thin at age 90 and harvest the bulk of the stand at age 100, leaving occasional clusters of seed trees. Seed trees are harvested at age 110, or after natural regeneration has begun. This process is called group selection. Following the yield table for 60 percent stocked stands of site index 70, approximately 7 Mbf per acre are harvested at age 100 and another 3 Mbf 10 years later. Costs under this system remain basically the same as previously indicated, except that artificial regeneration is unnecessary and thinning costs average $84 per acre.

The minimum stumpage prices necessary to offset management costs are displayed in table 7. Under the best of circumstances, unrealistically high stumpage prices must prevail. Prices must be even higher in this case than when timber management is guided by economic harvest criteria due to the longer time period over which administrative and protection costs accumulate and thinning costs appear. Timber management guided by physical harvest timing criteria

TABLE 7. Physical Criteria (Current Management) of Perpetual
Timber Management

Discount Rate of Minimum Stumpage Price (per Mbf)	Timber Primary Output (in dollars)	Multiple Outputs (in dollars)
7%	>1,000	39
10%	>10,000	45

fails the long-run efficiency test by an even greater margin than
timber management guided by economic harvest timing criteria.

Conclusion

We have examined public timber production, with emphasis on the
San Juan National Forest in southwestern Colorado. Our intent was to
test the hypothesis that public intervention in the timber market leads
to misallocation of funds in favor of expanded timber management
activities, and is also reflected in environmental destruction and
foregone recreational opportunity.

Railroad and mining demand combined with an abundant supply
of mature timber explain the historically viable timber market in this
region. Public agency intervention occurred in response to expected
long-run timber shortage. Today, it adds the justification of nonmarket values. That is, a public agency can provide for important social
values that free markets overlook. Since some of these values (e.g.,
wilderness recreation and protection of long-term environmental
productivity) conflict with timber, we expect their provision to imply
reduction of timber output to a level below that which would occur in
a free market.

Examination of Forest Service timber harvest and investment
criteria, however, suggests that the Forest Service applies more land
and capital to timber production than the free market would justify.
This conflicts with provision for some nonmarket values. Moreover,
the Forest Service fails to adjust land and capital investments for
timber in response to stumpage prices. Apparently, both justifications
for market intervention fail.

Our case study of both land the San Juan National Forest plans to
introduce into timber management and land it currently manages for
timber confirms this. A conservative estimate of identifiable initial
harvest costs for the first class of land is considerably in excess of

recent stumpage prices. Using economic harvest criteria, timber management costs also exceed recent stumpage prices for the higher quality currently managed land. This is true whether or not *compatible* nontimber forest resource uses absorb the common administrative and protection costs. Using the physical harvest criteria characteristic of Forest Service management, timber management costs exceed stumpage prices by an even greater margin.

We conclude that current production levels imply a large social cost. They are a drain on the federal treasury and they have a stumpage price decreasing effect that drives more efficient private timber producers from the market. The planned expansion of San Juan National Forest timber harvests would reinforce these negative effects. This proves our hypothesis. *Under current practice, more land is used than can be justified by a free timber market. Environmental destruction and foregone recreational opportunity result.* Reliance on the market would reduce timber production on environmentally risky sites. The reduction in timber land use increases the forest land available for wilderness recreation.

Environmentalists have appealed the expansion of timber management to roadless areas on the efficiency grounds just reviewed. Their appeal found favor with the regional forester; but the timber industry reappealed to the chief of the Forest Service in Washington, where the decision now rests.

We expect similar timber management problems elsewhere in Colorado and in the Rocky Mountains. Our conclusions are suggestive, but not restrictive, due to local variations in our empirical observations. Where efficiency argues for reduction in current harvest levels—as it does on the San Juan National Forest—the Forest Service resists on grounds of its responsibility to maintain community stability.

For example, the Beaverhead National Forest in western Montana plans to continue harvesting despite a Forest Service calculated benefit-cost ratio of less than one.[21] The justification is that local sawmills would go out of business and local people would be unemployed without Forest Service timber harvests. Local unemployment of labor and capital are one-time transition costs. It is difficult to understand how they can be used to justify perpetual inefficiency, particularly in the face of nontimber values that must be foregone. It is easier to understand the Forest Service responsibility to dependent communities in terms of easing the period of transition to reduce timber production levels. The transition period might reasonably be tied to local alternatives for the labor force and to labor mobility, as well as to the life of fixed capital investments. Moreover, application of efficiency criteria to all marginal public forest lands might reduce timber supply enough to create a price effect. The new, higher price would make some marginal lands efficient timber producers, thus partially offsetting predicted local unemployment.

Perhaps because of evidence like ours, the Forest Service is beginning to respond to efficiency criteria and to nonmarket values. Harvest levels have been reduced on some forests of the Southwest and the new regional forester in Denver is urging stricter justification of timber management. In both cases, there are obvious and immediate gains to environmental protection and recreational values. Most important, the Forest Service has initiated a functional accounting procedure that compares timber management costs directly with receipts. It will be difficult to implement because the Forest Service budget does not take this form, but, properly developed, it may provide a strong intra-agency incentive for economic efficiency in timber production and, concurrently, better provide for certain environmental and recreational values.

Notes

1. For a detailed explanation, see U.S. Department of Agriculture, Forest Service, "Forest Service Manual 2400: Timber Management," mimeographed (Washington, D.C., 1972); or William F. Hyde, *Timber Supply and Forestland Allocation* (Baltimore: Johns Hopkins University Press, 1979), appendix C.

2. Robert Marty and Walker Newman, "Opportunities for Timber Management Intensification on the National Forests," *Journal of Forestry* 67 (1969):482–85; Henry J. Baux, "How Much Land Do We Need for Timber Growing," *Journal of Forestry* 71 (1973):399–403; Marion Clawson and William F. Hyde, "Managing the Forests of Coastal Pacific Northwest for Maximum Social Returns," in *Timber Policy Issues in British Columbia,* ed. W. McKillop and W. J. Mean (Vancouver: University of British Columbia Press, 1976); Kurt Kutay, "Economic Impact Assessment of Proposed Wilderness Legislation," in *Oregon Omnibus Wilderness Act,* publ. no. 95–42, pt. 2 (Hearings before the Subcommittee on Parks and Recreation of the Committee on Energy and Natural Resources, U.S. Senate, 1977), pp. 29–63; and Hyde, *Timber Supply and Forestland Allocation.*

3. William R. Bentley and Dennis E. Teeguarden, "Financial Maturity: A Theoretical Review," *Forest Science* 11 (1965):76–87.

4. Unless otherwise noted, the data in this and succeeding paragraphs are based on U.S. Department of Agriculture, Forest Service, "Final En-

vironmental Statement for Timber Management Plan for the San Juan National Forest" (Durango, Colorado, 1976).

5. U.S. Department of Agriculture, Forest Service, "Summary of Sales in Roadless Areas (From 5-Year Action Plan)," mimeographed (Denver, Colorado, Region 2, 1976).

6. U.S. Department of Agriculture, Forms 2400–17, 1971–76.

7. Notice that the Forest Service concept of marginality differs from the economic concept. Where costs exceed revenues, the marginal-access or marginal-logging lands are economically submarginal. We use the Forest Service classification throughout when referring to these lands.

8. John V. Krutilla and Otto Eckstein, *Multiple Purpose River Development* (Baltimore: Johns Hopkins University Press, 1958).

9. This is not to suggest that timber mining cannot be economically efficient, only that it is forbidden on public lands. As previously noted, timber mining was apparently a successful private activity in earlier times. It is still socially optimal where harvest receipts exceed the sum of sale administration, road, environmental (e.g., soil displacement), and opportunity (including nonmarket) costs.

10. At $37,000 per mile, the actual road cost is $44 per Mbf. See Forest Service, "Final Environmental Statement for the San Juan National Forest."

11. Conversation with Walter Werner, timber management staff, San Juan National Forest, August 1976.

12. Ibid.

13. Ibid.

14. These stands are more heavily stocked than those on marginal access lands.

15. Darius Adams, Richard W. Haynes, and David R. Darr, "Price Effects of Changes in National Forest Timber Flows," mimeographed (Corvallis, Oregon State University, n.d.).

16. John V. Krutilla and Anthony C. Fisher, *The Economics of Natural Environments* (Baltimore: Johns Hopkins University Press, 1975).

17. W. H. Meyer, "Yield of Even-Aged Stands of Ponderosa Pine," Department of Agriculture Technical Bulletin no. 630, 1938.

18. Stocking refers to the density of the timber in the stand. Twenty percent of commercial forest land on the San Juan is nonstocked and 28 percent is understocked. Planned annual harvest of 35 Mbf of mature timber (over 100 years of age) from 11,000 acres (3.2 Mbf per acre) when compared with the above schedule, suggests an average stocking of considerably less than 60 percent. There is no reason to expect that stocking will be greater for later rotations without the introduction of additional silvicultural investments.

19. Office of Management and Budget, Circular A-94, 1973.

20. Hyde, *Timber Supply and Forestland Allocation*.

21. U.S. Department of Agriculture, Forest Service, "Land Management Plan for the Beaverhead National Forest" (Washington, D.C., 1978).

Transgenerational Equity and Natural Resources: Or, Too Bad We Don't Have Coal Rangers

by John Baden and Richard L. Stroup

The Ascendence of the Issue of Equity

Increasing attention is being paid to matters of equity. Such attention has most commonly been found in those intellectual areas where we expect it, that is, in the fringe whose outer limits of sanity are demarcated by the *CoEvolution Quarterly*. Of greater potential interest, however, is that a substantial number of mainstream academics also consider the issue. Included in this category are K. Boulding, F. Hirsch (*Social Limits to Growth*), R. Nisbet (*Twilight of Authority*), and D. Bell (*Cultural Contradictions of Capitalism*).

In his presidential address to the Association for Social Economics, Professor Joseph McGuire of the University of California, Irvine, states that "structural changes have already affected our economic policies (if not our theories), and are already so significant that they will probably affect the future of our American society. These basic alternatives in the social fabric stem largely from the shift in the position of the concept of equity in our contemporary culture."[1] The authors of the new equity literature are quite disparate in their evaluation of the social implications of a push for greater equity. Irving

The authors wish to thank Clay La Force, Harold Demsetz, and the Workshop in Law and Economics, Department of Economics, UCLA, for their comments and for providing the opportunity to present this article. We also extend our appreciation to Armen Alchian, Ken Godwin, Henry Goldstein, Verne Green, Ken Lyons, Bill Mitchell, Lloyd Orr, and James Quirk for their comments and to Walter Thurman for invaluable editorial assistance. Having rejected certain comments, the authors absolve these reviewers of responsibility for any remaining errors.

Kristol criticizes the social costs of an "entitlement society," while those identified with the Union of Radical Political Economists are convinced of its virtues. All parties, however, seem to have one area of agreement: there is a strong push toward increasing equity of outcome rather than opportunity.

The arguments involving equity can be conducted at several levels. Robert Nozick can engage in a learned discourse (*Anarchy, State, and Utopia*) with his Harvard colleague, John Rawls (*A Theory of Justice*), with the Harvard logician W. V. Quine as referee. Unfortunately, perhaps Keynes's mutterings regarding the importance of ideas deny us the luxury of believing that such debates are, at worst, harmless. Ignore them we can, since most people would, after all, rather die than think when self-interest is not directly involved; but we probably ignore with at least a twinge of guilt.

Rather than ignore, we ask you to consider. Consider the issue of equity—of central importance for moral public policy, yet inherently subjective and nonamenable to formal analysis. Obviously, the equity issue pervades any public decision. And regardless of the true motivations, the mask of equity, at least, is critical to the operation of political entrepreneurs. Further, and of more immediate concern, a misunderstanding of the equity implications of alternative arrangements of property rights has led a substantial number of well-intended, equity-minded people to advocate policies whose outcomes violate their equity preferences. We would like to correct a few of these misunderstandings.

Property Rights to Resources and Intergenerational Equity

If the human enterprise is expected to continue for at least several generations, the question of equity clearly has temporal as well as current spatial and positioned application. If those concerned with policy analysis are to become increasingly concerned with issues of equity, there is no obvious reason to restrict this concern to one generation. Thus, we should consider transgenerational equity. In its pure form, you are asked to assume that no one would know into which generation he or she or anyone else would be born. Once placed behind the veil of ignorance, our key question becomes: Which assignment of property rights will produce the greater degree of intergenerational transfer, an assignment of private rights or one with collective rights assigned to a democratic government?

It is taken by many as an article of faith that we are running out of resources despite the compelling evidence of static or declining real prices for many natural resources. Certainly a perception of resource depletion is real, regardless of the facts, and it is perceptions, *not* facts, that influence policy. Hence, if we are interested in policy, we must consider the perceptions that underlie policy.

Given a belief that we are running out of resources, it follows that one should expect future generations to be seriously disadvantaged. Those with an unfortunate later birthdate will suffer as a result of conscious consumption decisions taken by their predecessors, decisions that violate intergenerational equity.

If transgenerational equity is to be a goal, then, it becomes necessary to distribute the *value* of resources across generations. Obviously, it would be inequitable to distribute the volume or mass equally, for utilization efficiency will surely change. As a simple example, an equal volume of timber now produces, due to higher productivity efficiency, a higher volume and value of products now than it did forty or even ten years ago. Thus, were we to be allocated the same biomass of timber as was allocated to the previous generation, we would, in terms of a simplistic notion of equity, be unfairly advantaged.

Due to increased capital accumulation, including information and human capital, we expect improvements in utilization of all resources. Under incentives that reward efficiency, this outcome should occur partly due to the fact that resources become increasingly scarce. In this as in other areas, however, we expect to encounter diminishing marginal returns. The gain from moving utilization of standing biomass from 30 percent to 60 percent is likely to be easier to attain than a move from 60 percent to 90 percent.

The great wealth of capital stock available today was generated by the savings and accumulation of past generations. Call it altruism or poorly planned self-interest, the result is the same: each generation is endowed with a continually growing stock of productive capital with which it can satisfy its consumption desires as it sees fit. The natural resource equity argument is that this enhancement of consumption options is purchased at too high a price in terms of raw materials and natural amenities. Indeed, it seems reasonable to consider a possible shift in the relative opportunities offered by capital accumulation and raw materials. It is at least possible that future generations would prefer present generations to bequeath them less additional capital and more natural resources. As the authors of the *Federalist Papers* understood so well, no one can be assumed to be the best judge of another's preferences. Hence, those in the future might want the option of developing the capital that they find most useful. Clearly, however, each generation's use of resources influences the welfare of those that follow.

It is a blunt fact that the present generation operating in a historical context establishes the rules regarding property rights with respect to resources. While there may be no logical way to apply a discount rate for the comparison of satisfactions among different generations, each generation implicitly does so.

The Coase theorem tells us that with clear property rights the market mechanism will arrive at a Pareto optimum, even with externalities, provided that all parties can enter the market and that negotiations have negligible costs. But because future generations cannot bargain directly with the present, the Coasian approach is questioned.

Both the issues and the conditions should now be clear. Many consider equity to be increasingly important. Transgenerational equity (discounted by the probability of there being future generations) is one important form of equity. Property rights to resources are a component in an equity formulation. And, finally, future generations cannot speak for themselves.

The transgenerational equity questions may be stated quite simply. If one did not know into which generation he would be born, how would he structure property rights to resources? In the following sections we will undertake a preliminary analysis that turns out to yield counter-intuitive results.

Property Rights and Transgenerational Equity:
The Case of Stock Resources

We would all expect that a market system involving privately held rights would yield very different results than would a system whose rights were held by society and whose decisions regarding resource use were made collectively. It is widely believed that a market setting causes future generations to be robbed of natural resources. Krutilla and Page, for example, put it this way:

> Generally, markets are considered fair only if all those affected by the outcomes are present in the market (without externalities) and the distribution of market power is considered fair. In the case of deciding which new (energy) supplies to develop, the distribution of market power is indeed uneven: the present gen-

eration controls the total stock of resources, leaving future generations with no voice in today's decision.[2]

Further, Lippit and Hamada argue that, "In the extreme case, future generations cannot compensate the present for foregoing the mildest satisfactions, even when the very survival of mankind is at stake."[3]

The major implication of this and similar material is that a market mechanism, as compared with collective control, results in future generations being deprived of resources. But this claim does not withstand examination. The explanation results from both the different incentives faced by decision makers in the two situations and from the different ways decision makers are chosen in the two settings.

In what follows, simple models of market and collective democratic actions are employed. For concreteness, the resource stock in question will be referred to as a copper mine.[4] This example is chosen to capture the elements of intertemporal resource allocation and intergenerational transfer of resources, while presumably minimizing the intrusion of side issues: environmental externalities and violation of the exclusion principle. A binary decision must be made periodically on whether to exploit the one ore body in the current period or not. Following the initial analysis, the models will be made less naive by relaxing certain assumptions, and the resulting impacts will be noted.

To decide whether or not an existing resource should be exploited in the current time period, the decision maker simply compares its value (net of development costs) in current exploitation with its expected value in highest future use (net of development costs and discounted to the present). If current exploitation yields more net benefits, as judged by the decision maker, than does any future use, then the decision maker chooses current exploitation rather than preservation. The major difficulty, of course, lies in the estimation of value in future use. The value of a body of copper ore to be mined in any given future period depends on several factors, all of which are subject to uncertainty. Availability of other copper ore, the price of copper substitutes, the state of tastes and technology determining copper's usefulness and costs of development, attitude toward risk, and rate of discount are all important in determining a decision maker's estimate of the mine's present value in future exploitation. For a given mine, people are likely to have differing opinions on when the mine should be developed, or more specifically for present purposes, whether or not current exploitation is best.

The views of the populace on the present discounted value of future use might be summarized as in figure 6. The abscissa indicates $E(PV)$, the estimated present value of preservation, which is a single value in dollar terms, expressing the summation of the influences listed above. The ordinate indicates the frequency of each estimate. For simple models, no particular shape is required of the distribution. If we then locate on the abscissa a value, M, equal to the value (net of operating costs) of the ore body if mined now,[5] all $E(PV)$ greater than the value can be characterized as indicating that preservation is preferred. Similarly, those whose $E(PV)$ falls short of M, presumably must conclude that current development is the better choice.

Consider now the most straightforward kind of democratic decision making regarding the copper mine. Each person expresses his opinion of whether the mine should or should not be developed currently, and the majority rules. Assume that each individual is not simply self-interested, but that he votes for what he believes will benefit society most. To predict the outcome of such a vote, we simply ask whether the majority of the estimates fall to the right or to the left of the value of the mine in current use. If the majority is to the left,

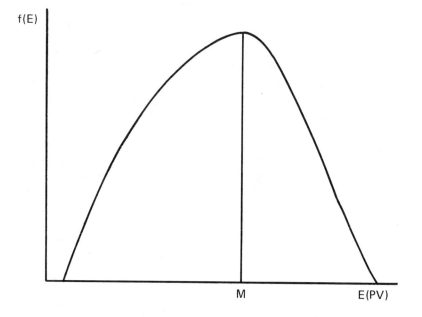

Fig. 6. Expected values of preservation versus current exploitation

current exploitation will be mandated; if to the right, preservation is supported. Put another way, if the median voter[6] has $E(PV)$ greater than M, preservation will result, while current development wins if he feels the other way. In a very real sense, the median voter's judgment prevails.

Alternatively, consider a simple market situation involving the same people with the same tastes, expectations, and discount rates, where the copper mine is controlled by the highest bidder. One type of bid is M, for current development, made on behalf of ore processors. The highest such bid represents the mine's worth in current exploitation. The other type of bid is from those who want to preserve the mine for the future. We can assume either altruistic or selfish motives for these bidders. In either case, each bid reflects the bidder's belief as to the mine's value. Obviously, if anyone (with sufficient funds or credit or the ability to convince coventurers) believes the mine will be sufficiently more valuable in the future than it is now, so as to justify postponing its use, the resource will be preserved. The median opinion does not control. The tendency instead is for those *with the strongest preservation bias* to control. They are usually called speculators.

We have long been puzzled regarding the general condemnation of speculators by environmentalists and preservationists. Speculator is, quite widely, a derisive term. Yet, such criticism seems to be at variance with the announced preferences of the critics. The critics whom we address claim to favor deferred resource consumption, which is merely saving it for the future. This, of course, is exactly the function of the speculator. Only by paying a higher price than those who prefer to consume now can he conserve the resource for his profit (and for the future). While current consumers have good reason to object to speculators for driving up the price and hence reducing current consumption, those in the future should shower them with praise and rewards—if the speculator guessed correctly. The central point, of course, is that successful speculators advantage consumers in the future at the expense of those in the present. Their action in markets over time is analogous to distributors of goods over space. The distributor of oranges buys in Florida on behalf of New Yorkers. Florida orange prices would be lower (and Florida consumption of them higher) if interstate trade were forbidden.

Whether the speculators have a long view encompassing the future period when the resource will be developed or a more short-sighted view for their own personal financial plans is unimportant. So long as they can transfer (sell) the property rights they hold on the mine, and so long as a few others hold a similar view of the mine's future value, the mine remains a salable asset and a good investment. As time passes and the higher valued time of use approaches, the present discounted value rises.

Of course, if the purchasing speculator is wrong and potential bidders begin to learn so, he suffers the loss as the mine's value rises less rapidly (or falls) compared to other assets he could have held. He *and* the deprived earlier generation bear society's loss if his decision is incorrect. *But the resource is preserved.* Since this type of speculative activity can be expected whenever resource property rights are private and transferable, resource prices in such markets will reflect bidding for future use, and *current exploitation will occur only when all speculative bids are overcome.* Contrary to the statements by Krutilla and Page, the equilibrium market price clearly includes pressure from future potential bidders, including those bidders yet unborn, since speculative bids are based on what *future* users are expected to be willing to pay. Hence, in a market system with transferable property rights over stock resources, those who are *most optimistic* regarding the *future* value of any storable are the ones who control the resource. Given that they believe that the future value will be high, they expect to capture rewards by keeping resources out of consumption.

It is difficult to imagine how a mechanism could be devised to give current voters an analogous incentive to consider future citizens. Future voters must depend on the good will of present voters to sacrifice current consumption of governmentally controlled resources. Our analysis of collective control has thus far assumed that such good will is present—that present voters view future generations' consumption as they do their own. The only discount factor assumed to apply to consumption in the distant future was that which people apply to their own consumption during their lifetimes. This form of altruism was not required of the private bidders.

If we allow more self-interested voters to enter our collective control model, the preservation bias of the market stands out in even sharper relief. If voters are less interested in future generations' welfare than in their own, current exploitation becomes more valuable relative to the benefits of preservation in the eyes of current voters. The value in current use, M, remains constant while their effective $E(PV)$ falls because future usefulness, enjoyed by others, is more heavily discussed than if current voters themselves could enjoy the benefits.

It should be clear that as we allow for self-interested behavior, the most realistic presumption is not that voters feel toward future generations as they do their heirs. It can be argued (particularly well in sociobiological terms) that such a presumption collapses back to the naive altruistic view. People in general may value their descendants' consumption as they do their own. However, the voters deciding on the stock of natural resources to bequeath to the next generation are not considering descendants' welfare alone, but the welfare of all those alive in the future. Such a diffused interest will surely result in a lower present value than that which leads people individually to leave bequests to their heirs. On the other hand, since costs also are diffuse, the net effect is not obvious.

Another assumption to be relaxed is that of market structure in the private control model. Initially, we posited a competitive bidding process for the resource. In fact, a competitive market is not necessary to our results. In a monopolized or cartelized market, the tendency toward preservation is increased. As Hotelling demonstrated in 1931, a constant-cost monopoly will restrict the exploitation rate due to its output-restricting behavior.[7]

To summarize the situation with stock resources, privately held, exchangeable property rights tend to *encourage* preservation, relative to a simple democratically controlled collective management system. This is because the gains from preservation are appropriable in a market system, but not with collective ownership, and because those with expectations of high future value for the resource tend systematically to control through outbidding others. The preservation bias differential is increased if people are viewed as self-interested, or if the private producing industry is a monopoly or a cartel.

One implication of the stock resource model is at variance with commonly accepted wisdom. A respected source of that wisdom is Robert Solow, who in his 1973 Richard T. Ely Lecture states:

> We know in general that even well-functioning competitive markets may fail to allocate resources properly over time. The reason, I have suggested, is because, in the nature of the case, the future brings no endowment of its own to whatever markets actually exist.[8]

We have argued that, at least relative to collective control, the future does have a representative in present markets: the speculator. The endowment the future brings to the market is what the speculator expects the future to be willing to pay.

Solow suggests a partial corrective to the perceived lack of representation of the future. Futures markets are claimed to save resources for future generations.[9] Our analysis suggests the opposite. To institute a futures market is to allow speculators to be supplied, not only with actual claims on resources, but speculative claims as well. Without futures contracts, the only role for the speculator is to bet on the rises in resource values. Futures contracts allow speculators to sell short those resources they expect to decline in value, thereby depressing current prices and encouraging greater current consumption of those resources. In short, the futures market gives influence in the resource market to those expecting a lower rise in resource price, or having a higher prediction of the social discount rate.[10]

Transgenerational Equity and Flow Resources: Or, Too Bad We Don't Have Coal Rangers

We have argued that market decisions are expected to utilize nonrenewable stock resources more slowly than democratic collective decisions. In a market, those who expect resources to be especially valuable in the future tend to capture control over the resources and withhold them from present consumption. Here, self-interest is harnessed in the service of future generations.

When dealing with bureaucratically managed renewable or flow resources (such as timber), the incentives, and hence the outcome, are greatly different.[11] With nonrenewable resources, self-interested market behavior is expected to produce more conservation than the median voter would prefer. In contrast, collectively owned and collectively managed renewable resources are expected to be both (*a*) conserved more and (*b*) produced more input-intensively than they would be under private ownership. This is not, however, to suggest that collective ownership and management will result in socially optimal rates of use and investment. We merely state that collective ownership and management will *tend* to push benefits into the future. It may be that the pathologies inherent in this system are so severe that all generations lose from collective ownership, but the resolution of that question is outside the scope of this paper.

Unlike coal or copper, trees grow during a relevant planning horizon. Together with the need for management, this is the key to understanding the different outcomes under market and collective ownership. Growth is a process that can be influenced by intervention. Silvicultural treatments influence both composition and speed of timber growth. Hence, there is a basis for investments in forestry. To this point, the situation is very simple, and there are no grounds to predict whether markets or collective management will produce and conserve more for the future.

Two considerations complicate this issue. First is the equimarginal principle, which states that resources will be most efficiently deployed when the marginal return on units of investment in each opportunity are equal. Related to this is the principle of diminishing marginal returns. This means that for every investment opportunity there is a point at which the return from an additional unit of input is less than from the previous unit. The discipline of the market is such that the successful decision makers are those who act as though they pay attention to these two principles.[12] Hence, in the private sector, both standing inventory decisions and silvicultural investments are filtered through reality checks. Firms that consistently ignore the implications of these principles are called bankrupt. Thus, the market system operates to separate systematically unlucky, incompetent, or lazy people from the control of resources. Depending on one's ideology, this may be either good or bad; but the verity of this claim seems evident.

We are arguing that resource use is a function of the incentives faced by resource managers. In private markets with well-defined property rights, the incentives serve to maximize the value of output from flow resources or to minimize the value of inputs for a given output flow. Public managers are no different from private managers in that they tend to respond to incentives. Both are largely self-interested. McKenzie and Tullock have perhaps the classic statement:

> Bureaucrats are not markedly different from other people. Most citizens of the U.S. are to some extent interested in helping their fellowman and in doing things in the public interest. Most citizens of the U.S., on the other hand, tend to devote much more time and attention to their own personal interests. The same is true of bureaucrats.[13]

One explanation for more stocking and more production investments in collectively owned and bureaucratically managed forests is based on the incentive structure faced by the bureaucratic managers.

For people in general, but for highly motivated individuals in particular, self-interest leads to the desire for an increase in discretionary control over resources. For the "selfish" individual, this provides the power and deference that accompany discretionary control. For the professionally oriented or "socially concerned" individual, this provides in addition the ability to make "good" things happen. More timber growth is presumably a good thing to a forester, for example. When resources are owned collectively, as in a bureaucracy such as the Forest Service, a prime strategy for increased discretion is to promote growth of one's bureau.

There are reasons to believe that, in most cases, waste is generated when the bureau is above optimum size. Most will agree that substantial forces lead in this direction. For the bureau head, his civil service rank, prestige, and pay are all strongly related to bureau size. Further, symbols of success in terms of office amenities are also related to the number of persons under his charge (for example, in one university, for years only deans and above could have IBM typewriters). In addition, expansion generates more possibilities for promotion. This enhances the ability to control those under his charge, since under Civil Service rules firings are nearly impossible to execute successfully. Thus, to gain control over inferiors, the promise of promotions may be offered as inducements. And promotions are more common in a situation of growth.

Of perhaps equal importance for the ambitious bureaucrat is the fact that a large proportion of his budget is "locked in" from previous years. This, of course, reduces the range of discretionary expenditures. In contrast, new funds offer far more opportunities for flexibility and for innovation.

Among other results, this tendency toward growth can be expected to encourage decisions leading to more intensive management of this resource and the reluctance to surrender territorial authority (unless the cut in manpower is small or exercise of the authority leaves no *discretionary* resource claims) or to merge with any larger entity or to transfer resources to activities outside the agency's scope. Such incentives are consistent with maximum preservation of the resource or large (relative to private) inventories.

This strong desire for growth does not depend on the presence of evil administrators or megalomaniacs. We must remember that the bureaucrat, because he lacks market information on the relative value of his product and those of other public agencies, suffers from the absence of an obvious and immediate reality check on what he wants to believe. Thus, it is easy for him to harbor the illusion that his agency mission is of above average merit and thus argue that his office deserves above average budget increases. He, of course, has the help of clientele groups at budget time. Collective ownership and the lack of a pricing mechanism result in both antiefficient incentives and distorted information—or lack of the latter—which deal blows to even a well-meaning, intelligent bureaucrat from which recovery is difficult and rare. In sum, the bias is toward expanded activity. When dealing with renewable resources, this means high flows and high inventories, since there is no interest charged to the inventories. A large resource stock, to the extent that it justifies larger bureaucratic budgets, will be favored by those in charge of the bureaucracy.

Conclusion

If there is increasing concern with equity, this concern can reasonably include transgenerational equity. If one were ignorant as to the generation into which he would be born, he would advocate some mix of property rights over resources. If nonrenewable stock resources are to be saved for the future, transferable property rights should be established over them. When considering renewable resources, one concerned about having a large inventory and ignorant regarding the time of his birth might advocate taking advantage of bureaucratic pathologies and putting these resources into collective ownership and bureaucratic management. If, as some have advocated, nonrenewable stock resources are to be put into public ownership, then those in the future might suggest that we build schools for the training of coal rangers. This would appear to be far less wasteful of resources than many governmental activities undertaken in natural resource fields.

Notes

1. J. McGuire, *Review of Social Economics* 33 (April 1975):74–80.

2. J. V. Krutilla and R. T. Page, "Paying Tomorrow for Energy Today," *Resources* 49 (June 1975).

3. V. D. Lippit and K. Hamad, "Efficiency and Equity in Intergenerational Distribution," in *Sustainable Society*, ed. Dennis C. Pirages (New York: Praeger, 1977), pp. 285–99.

4. Production from the copper mine other than copper ore will be ignored for the sake of simplicity.

5. In reality, the value of the mine in "current" development is subject to some uncertainty, particularly since development is not really confined to one short time period. But the degree of uncertainty is small relative to development farther into the future. One could work with a similar, though much compressed, distribution of estimates of value in current use, but that

would seem to add complexity with no change in the basic outcome in comparing private and collective management systems.

6. Median voter is the individual whose $E(P)$ splits the distribution; half the people lie above him and half below.

7. H. Hotelling, "The Economics of Exhaustible Resources," *Journal of Political Economy* 39 (April 1931).

8. R. Solow, "The Economics of Resources or the Resources of Economics," *American Economic Review* 64 (May 1974):10.

9. Ibid., p. 13.

10. Note that this same reasoning explains why farmers might dislike futures trading in their commodities. Claims to future supplies are sold in competition with producer supplies. No seller wants sales competition.

11. The important variable here is whether management intensity is correlated with larger stock, as usually is the case in renewable resources, or with more rapid exploitation, as with nonrenewable resources.

12. If confused by this, review Milton Friedman's *Essays in Positive Economics* (Chicago: University of Chicago Press, 1966), chap. 1.

13. R. McKenzie and G. Tullock, *The New World of Economics* (Homewood, Ill.: Richard D. Irwin, 1975).

The Environmental Costs of Bureaucratic Governance: Theory and Cases

by M. Bruce Johnson

During the initial stages of the recent debate on the environment, some critics of the private sector believed that self-interest, private property, the profit motive, and markets inherently foster environmental destruction. Their solution was to transfer property rights and control of resources to government bureaucrats who were not property owners or residual claimants and, hence, would be dedicated to promoting the public interest.

The essays in this volume suggest that the public interest strategy has not lived up to its promise. As well-intentioned as it may have been, the transfer of resource control from private ownership and markets to public ownership and governmental bureaucratic control has not been the panacea some expected. The evidence suggests that, traditional issues aside, governmental ownership or control has produced nonmarket environmental costs rather than benefits on balance.

What went wrong? The answer is not that we need "better" people in the bureaucracy. To chastise the responsible bureaucrats for incompetence or bad faith is just as superficial as to condemn private resource owners in the market. Instead, the disappointing performance of governmental ownership and management of natural resources results from the faulty institutional arrangements within which resource decisions are made. Government bureaucrats, as individuals, have the same goals as the rest of us. They want higher incomes, promotions, discretionary authority, security, and a variety of work place amenities. They will promote these goals in the context of the institutional setting in which they find themselves. If they are free from the discipline of the market, they will ignore the prices the market would otherwise place on resources. If the institutional arrangement permits bureaucrats to tap the general budget, the incentives inherent in the institutional structure will permit them to rationally regard the budget as a common pool resource and to exploit it. This argument is developed in this volume by Terry L. Anderson and Peter J. Hill, and by Rodney D. Fort and John A. Baden. Similarly, bureaucrats are encouraged to transfer benefits to well-identified groups and spread the costs among taxpayers without regard for the effects these transfers will have on efficient resource use.

By way of background and clarification, it is worth reflecting on the rationale that led to our current disappointments with public resource ownership and management. With our tradition of pragmatic activism, we Americans are compulsive in our willingness to tackle tough social problems. Whenever we perceive a problem or an injustice, we can count on the emergence of a hero, a heroine, or a coalition to propose some way to fix it.

As predictable as these heroic responses are, they have traditionally been preoccupied with results rather than processes. Public spirited reactions to the environment, inflation, unemployment, energy, and all the other issues of the day are *result* rather than *process* oriented. If we think too many trees are being cut, we simply legislate that fewer trees be cut. If we are disturbed by inflation, we propose the adoption of price controls. If we are concerned by unemployment rates, we propose that government create additional jobs. In each case, we focus directly on the desired result rather than on the processes, mechanisms, and institutions that might reasonably be expected to lead to a preferred condition.

Enter the economist as the curmudgeon with the message that good intentions are not enough to achieve desirable ends. It is the process that counts. Processes are not only important, they are critical to producing desired results. Since the economic and political landscape is littered with the wreckage of well-intentioned but disappointing programs, the thoughtful activist cannot ignore the economists' warnings. Government environmental protection programs have not fulfilled their positive promises on the one hand and have led to unanticipated negative consequences on the other. Enough evidence is now available to suggest that the resulting frustration cannot be eliminated by means of "better" programs with "better" people running them. Instead, it is obvious that more attention should be devoted to the institutions and processes that led to the original, undesirable outcome as well as to the processes set in motion when we adopt new programs to solve the problems.

The economist's intellectual perspective on these matters is so similar to the environmentalist's that one wonders why a partnership between the two was not formed in the natural course of events. Environmentalists believe that the ecological system is a collection of interconnected and interdependent parts. An exogenous change in one sector or subsystem will lead to a chain reaction of effects elsewhere. Economists hold the identical view with respect to the economic system. General equilibrium models are a formal way of saying that everything depends on everything else. Moreover, it is axiomatic in economics that individuals make choices with some specific purpose in mind. Hence, given different opportunities, individuals will respond in different ways. Perhaps more to the point, prices and incentives *do* make a difference to the outcome. The lesson for environmental policy is simply that the programs used, however well intentioned, will fail if the incentives buried in the institutional structure are ignored or poorly designed.

What difference does this make? A great deal in both theory and practice. An environmental improvement program that neglects its effects on the incentives of interested parties will lead to unintended consequences. Some of the essays in this volume document that result in specific cases. For example, William F. Hyde finds that the Forest Service's definition of "maturity" leads to excessive investment in capital in older and larger stands of timber. Because the Forest Service ignores prices and interest rates in most of its decision making, it extends timber growing into marginal lands where the rates of return are negative. In some cases, the cost of timber harvesting exceeds the gross monetary value of the timber sold and, in addition, environmental values are sacrificed. Sabine Kremp suggests that the Bureau of Land Management's blanket policy of specialized grazing management is both economically inefficient and environmentally damaging. Bernard Shanks finds that government water projects typically sidestep objective cost-benefit tests. He places the blame on the failure of our institutions and processes to separate the responsibility for project justification from the responsibility for construction and management.

Gary D. Libecap and Ronald N. Johnson argue that overgrazing on the Nevajo reservation is the result of the failure by the Department of the Interior and the Tribal Council to enforce private property rights in land. Both agencies follow policies that break up large herds into small ones and thereby raise the costs of negotiating, transferring, and enforcing property rights. Ronald M. Lanner identifies the environmental costs and benefits of government chaining of piñon and juniper, as governmental agencies arbitrarily promote the production of cattle on range land as opposed to environmental amenities on noncommercial forest land. Barney Dowdle points to the sustained yield-even flow theory as an institutional anachronism that wastes the public timber wealth and the taxpayers' dollar. Persistent adherence to the Forest Service's even flow criteria forces society inside its production possibility curve for both wood and environmental benefits and, not insignificantly, encourages timber processing firms to invest in nonoptimal amounts and types of capacity.

Ernst Habicht, Jr., demonstrates how natural gas policies have resulted in an understatement of the true social costs of natural gas and, consequently, have encouraged the overuse of natural gas and discouraged industrial and commercial users from switching to alternative fuels. Tax codes and utility regulatory policies reward investment in conventional energy technologies and suppliers or, occasionally, in very expensive alternatives like synthetic natural gas, but then regardless of the cost. In a similar vein, Richard L. Stroup argues that the current rationale for synthetic natural gas, expensive and environmentally damaging as it is, is the result of the natural gas shortage caused by the regulation of wellhead natural gas prices. If price controls were lifted, the market price of natural gas would preclude synthetic natural gas as a cost-effective alternative in the near future.

Lloyd D. Orr investigates the social costs of the incentive structure inherent in current environmental policies and finds that these policies create excessive costs, destroy private property values unnecessarily and arbitrarily, and inhibit the operation of the decentralized decision structure that is essential to establishing and maintaining a high quality environment. In particular, he argues that the strategy of effluent charges is a vastly superior method for promulgating sound environmental policies. John Baden and Richard L. Stroup address some of the alternative methods of resource allocation and the associated implications for inter-generational equity. They argue that collective ownership involves the lack of appropriate pricing mechanisms and results in distorted information, a bias toward overactivity, and incentives that are per se inefficient.

Thus, in a wide variety of cases, governmental action leads to a net waste of resources *and* to a degradation of the environment. How can this behavior be explained except by a perverse incentive structure? Assuming that government bureaucrats want neither to waste resources nor to spoil the environment, their behavior must be a purposeful, reasoned response to the incentive structures in which they find themselves operating. It must be in their best interest to act as they do.

But the failure of environmental programs cannot be blamed wholly on the bureaucracy. Various segments of the public are also systematically advantaged or disadvantaged by environmental programs, and these individuals and groups naturally adjust to the new programs in order to either make the best of a bad situation, cut their losses, or beat the system. That, too, is rational behavior that an economist would trace to the incentive structures and the prices implicit in the environmental program. Again, it is surprising that the environmentalist sponsors of a program would not anticipate the feedback and interdependence in the political and economic realm with the same degree of insight that they have in the ecological realm. This natural, purposeful behavior by affected parties—whether bureaucrats or otherwise—subverts the programs.

In a more general sense, the environmentalists frequently confuse normative and positive issues. To be specific, since many environmentalists believe that environmental amenities *should* be free to all, they support and promote programs intended to produce environmental amenities free of user charges. Here the economic distinction between price and cost is relevant. The program may indeed set the *prices* of environmental goods at zero, but the true *costs* incurred by society are unambiguously positive! These costs are measured in terms of the foregone alternatives. For example, if a commercially viable forest of any size is set aside for aesthetic purposes, the cost of the amenities thus produced can be measured in terms of higher timber prices in general and in lower receipts, taxes, and employment in the immediate vicinity. The fact that environmental amenities associated with an untouched forest are assigned a zero price (i.e., given away free) only masks the real costs of the program and deludes the public. Furthermore, when the prices of environmental goods are set at zero (or, at least, set much lower than the true costs), a certain dynamic process is set in motion. The public believes the environmental goods are free and, as a consequence, rationally demands that still more goods be produced. Why not, if it is free?

This pricing policy and the resulting dynamic are not unique to environmental programs. Because of the institutional structure, government and its special interest allies will systematically underprice the output of their programs on a case-by-case basis. The systematic understatement of prices of environmental and other goods and services produced by government has far-reaching implications in an interdependent and interconnected economic system. Most people rely on prices as an efficient source of information to guide their purchases. When a typical consumer buys a phonograph record in the private marketplace, he does not have to become informed about the price of the appropriate plastics, pressing equipment, delivery and packaging costs, and so forth. All of that information is summarized in the price of the record. In the case of government goods, the typical consumer/taxpayer also looks to the "price" as his source of information. If the government quotes a price lower than the true cost of the goods or services, individual consumers and groups can be expected to want more of the governmental output than would be the case if the higher, correct price were quoted.

This conclusion is critical. In the absence of a market price, who among the general public really knows the true costs of any good—especially environmental goods? If the government does not assign the correct price to the output, can the individual be reasonably expected to engage in laborious and intricate cost-benefit analyses? Moreover, the citizen/taxpayer with a special interest axe to grind is encouraged to engage in strategic behavior. He correctly reasons that the cost of his special program will be divided among the general public and, consequently, his personal benefit-cost ratio will be high. This is similar to the situation in which ten people go out to dinner and agree before hand to divide the total check for the group equally, even though each individual can order whatever he or she pleases. If a dinner is a one-time affair with strangers, the rational diner will reason as follows: "If we all order a $10.00 dinner, the total check will come to $100.00 and I will pay $10.00. But, if I order a $20.00 dinner while the other nine order the $10.00 dinner, the total check will be $110.00 and we will each pay $11.00. Thus, I will get a $20.00 dinner for $11.00." Extend the logic to the next step. Since each diner has an incentive to splurge, can there be little doubt that the total amount of money spent on the dinner will be greater under the "equal share" agreement than under an arrangement where each diner paid his or her own check?

The analogy to the production of government goods financed out of the general tax fund is clear. Each special interest group will lobby to increase the activity level of its own pet projects on the rational expectation that the bulk of the costs will be passed on to the larger taxpaying public. The net result is a larger public expenditure and greater governmental activity than each citizen would freely choose in the absence of this perverse incentive structure. Adopting user charges for most, if not all, governmentally produced goods and services would internalize the costs and produce both better levels and better mixes of output. Failing that, governmental activity must be directed by distributional or equity considerations that are arbitrary, at best.

Perhaps more to the point, the production of "free" public goods leads to strategic gaming behavior by citizens and to the disappointments and inefficiencies so clearly documented in this volume. The axiom, "There is no such thing as a free lunch," also extends to vistas, natural gas, air quality, water projects, wilderness areas, and grazing land. Society will indeed pay the price of the lunch. The accumulating evidence suggests that the price will be in terms of less output *and* a worsened environment.

Selected Bibliography

Alchian, Armen, and Demsetz, Harold. "Property Rights Paradigm." *Journal of Economic History* 33 (March 1973).

Anderson, Terry L., and Hill, P. J. "The Evolution of Property Rights: A Study of the American West." *Journal of Law and Economics* 18 (April 1975):163–79.

————. "Toward a General Theory of Institutional Change." *Frontiers of Economics.* Blacksburg, Va.: University Publications, 1976, pp. 3–18.

————. *The Birth of a Transfer Society.* Stanford, Calif.: Hoover Institution Press, 1980.

Baden, John A., and Stroup, Richard L. "Private Rights, Public Choices, and the Management of National Forests." *Western Wildlands,* Autumn 1975, pp. 5–13.

————. "The Environmental Costs of Government Action." *Policy Review,* Spring 1978, pp. 23–36.

Berkman, Richard L., and Viscusi, W. Kip. *Damming the West.* New York: Grossman, 1973.

Borcherding, T. E., ed. *Budgets and Bureaucrats: The Sources of Government Growth.* Durham, N.C.: Duke University Press, 1977.

Breyer, S. G., and MacAvoy, P. W. *Energy Regulation by the Federal Power Commission.* Washington, D.C.: The Brookings Institution, 1974.

Cheung, Steven N. S. "The Structure of a Contract and the Theory of a Non-Exclusive Resource." *Journal of Law and Economics* 13 (1970):49–70.

Clawson, Marion. "The National Forests." *Science* 191 (1976):762–67.

Dales, John. *Pollution, Property, and Prices.* Toronto: University of Toronto Press, 1968.

David, P. A. *Technical Choice, Innovation and Economic Growth.* New York: Cambridge University Press, 1975.

Davis, Kenneth P. *Forest Management.* New York: McGraw-Hill, 1966.

Demsetz, Harold. "Some Aspects of Property Rights." *Journal of Law and Economics* 9 (October 1966):61–70.

————. "Toward a Theory of Property Rights." *American Economic Review* 57 (May 1967):347–59.

Dennen, R. Taylor. "Cattlemen's Associations and Property Rights in Land in the American West." *Explorations in Economic History* 13 (October 1976).

————. "Some Efficiency Effects of Nineteenth-Century Federal Land Policy: A Dynamic Analysis." *Agricultural History* 51 (October 1977).

Driscoll, R. S. *Managing Public Rangelands: Effective Livestock Grazing Practices and Systems for National Forests and National Grasslands.* Washington, D.C.: Government Printing Office, 1967.

Gardner, D. B. "Transfer Restrictions and Misallocation in Grazing Public Range." *Journal of Farm Economics* 44 (1962):109–20.

Gwartney, J. D., and Stroup, R. L. *Economics: Private and Public Choice.* 2d ed. New York: Academic Press, 1980.

Habicht, E. R., Jr. "Electric Utilities and Solar Energy: Competition, Subsidies, Ownership and Prices." In *The Solar Market: Proceedings of the Symposium on Competition in the Solar Energy Industry.* Washington, D.C.: Federal Trade Commission, 1978.

Hardin, Garrett, and Baden, John. *Managing the Commons.* San Francisco: Freeman, 1977.

Haveman, Robert. *The Economic Performance of Public Investments.* Baltimore: Johns Hopkins University Press, 1972.

Holling, C. S., and Goldberg, M. A. "Ecology and Planning." *American Institute of Planners Journal* 37 (July 1971):221–30.

Hormay, A. L. *Principles of Rest-Rotation Grazing and Multiple Use Land Management.* Washington, D.C.: Government Printing Office, 1970.

Hotelling, H. "The Economics of Exhaustible Resources." *Journal of Political Economy* 39 (April 1931).

Hyde, William F. *Timber Supply and Forestland Allocation.* Baltimore: Johns Hopkins University Press, 1979.

Kneese, A., and Bower, B. *Managing Water Quality: Economics, Technology, Institutions.* Baltimore: Johns Hopkins University Press, 1968.

Kneese, A., and Schultze, C. *Pollution, Prices and Public Policy.* Washington, D.C.: The Brookings Institution, 1975.

Krutilla, John V., and Fisher, Anthony C. *The Economics of Natural Environments.* Baltimore: Johns Hopkins University Press, 1975.

Libecap, Gary D. "Economic Variables and the Development of the Law: The Case of Western Mineral Rights." *Journal of Economic History* 38 (June 1978).

McKenzie, R. B., and Tullock, G. *Modern Political Economy.* New York: McGraw-Hill, 1978.

Mills, Edwin. *The Economics of Environmental Quality.* New York: W. W. Norton, 1978.

Niskanen, W. A., Jr. *Bureaucracy and Representative Government.* Chicago: Aldine-Atherton, 1971.

Pirages, Dennis C. *Sustainable Society.* New York: Praeger, 1977.

Public Land Law Review Commission. *One Third of the Nation's Land.* Washington, D.C.: Government Printing Office, 1970.

Ridgeway, Marian E. *The Missouri Basin's Pick-Sloan Plan.* Illinois Studies in the Social Sciences, vol. 35. Champaign: University of Illinois Press, 1955.

Rosenberg, N. *Technology and American Economic Growth.* New York: M. E. Sharpe, 1972.

Rourke, Francis E. *Bureaucracy, Politics, and Public Policy.* New York: Little, Brown, & Co., 1976.

Schultze, C. *The Public Use of Private Interest.* Washington, D.C.: The Brookings Institution, 1977.

Shiflet, T. N., and Heady, H. F. *Specialized Grazing Systems: Their Place in Range Management.* Washington, D.C.: Government Printing Office, 1971.

Solow, R. "The Economics of Resources or the Resources of Economics." *American Economic Review* 64 (May 1974):1–14.

Stroup, R. L., and Baden, J. A. "Externality, Property Rights, and the Management of Our National Forests." *Journal of Law and Economics* 16 (Spring 1973).

Stroup, R. L., and Thurman, W. N. "Will Coal Gasification Come to the Northern Great Plains?" *Montana Business Quarterly* 14 (Winter 1976):33–39.

U.S. Department of Agriculture, Economic Research Service. *History of Federal Water Resources Programs, 1800–1960.* Misc. publications no. 1233. Washington, D.C.: Government Printing Office, 1972.

U.S. Department of Interior. *The Public Lands.* Washington, D.C.: Government Printing Office, 1966.

————, Office of Coal Research. *Evaluation of Coal Gasification Technology. Part I: Pipeline Quality Gas.* Washington, D.C.: Government Printing Office, 1973.

Contributors

TERRY L. ANDERSON	Associate, Center for Political Economy and Natural Resources, Montana State University
JOHN BADEN	Director, Center for Political Economy and Natural Resources, Montana State University
BARNEY DOWDLE	Department of Forestry, University of Washington
RODNEY D. FORT	Graduate research assistant, Center for Political Economy and Natural Resources, Montana State University
ERNST R. HABICHT, JR.	Independent consultant, Port Jefferson, New York; former chairman, Energy Program, Environmental Defense Fund
PETER J. HILL	Associate, Center for Political Economy and Natural Resources, Montana State University
WILLIAM F. HYDE	Center for Resource and Environmental Policy Research, Duke University
M. BRUCE JOHNSON	Department of Economics, University of California, Santa Barbara
RONALD N. JOHNSON	Department of Economics, University of New Mexico
SABINE KREMP	Forest Service
RONALD M. LANNER	Department of Forestry and Outdoor Recreation, Utah State University
GARY D. LIBECAP	Department of Economics, University of New Mexico
LLOYD D. ORR	Department of Economics, Indiana University
BERNARD SHANKS	Department of Forest Science, Utah State University; Wilderness Society
RICHARD L. STROUP	Codirector, Center for Political Economy and Natural Resources, Montana State University

Index

Accountability, 7, 180
Alaska Purchase of 1867, 125
Allowable cut effect (ACE), 175, 176, 177, 178, 183, 185 n.11
Altruism, 5, 71, 127, 205, 209, 210–11
Amenities, 5, 146, 181, 183, 196, 205, 214, 217, 220, 221
American Revolution, 29, 65, 125
Annual allowable cut, 174
Antiquities Act, 162
Appropriations Committee, United States Congress, 13, 14, 129
Army Corps of Engineers, 108, 110, 111, 112, 114, 116, 118. *See also* Water development agencies
Associations, voluntary, 23, 30–31, 32, 33, 40, 41, 42, 125
Auctions, 29, 34, 43 n.9, 144, 182, 190, 209, 210
Austin, T. Louis, 70

Bargaining, 26, 30, 55, 56, 58, 206
Benefit-cost analysis, 17, 55, 57, 100, 109, 113, 114, 115, 116, 118, 121, 164–65, 176–77, 219, 222
Benefits, 4, 38, 51, 108, 112, 113, 115, 119, 120, 179, 197, 218, 220
Bounties, 65
Brandeis, Louis D., 38
Budgets, 5, 13, 16, 17, 176, 179, 218; as common pool, 15, 218; competition for, 14, 15; distribution of, 14; increases in, 178, 214; reduction in, 18, 19; scarcity of, 14
Bureau of Budgetary Control (BBC), 19, 20

Bureau of Indian Affairs (BIA), 85, 88, 100, 102. *See also* Indian Service
Bureau of Land Management (BLM), 6, 124, 128, 129, 130, 146, 147, 156, 165, 167, 219; grazing systems of, 142; legal action against, 163; management objective of, 124, 130, 133, 143; responsibilities of, 128. *See also* General Land Office
Bureau of Reclamation, 19, 109, 110, 111, 112, 114. *See also* Water development agencies
Bureaucracy, 2, 9, 13, 18, 20, 54, 214, 221
Bureaucratic pathology, 17, 18, 19, 121
Bureaucrats, 1, 4, 5, 13, 14, 15, 69, 70, 121, 146, 212, 213, 215, 217, 218, 220, 221; activities of, 12; and budget, 214, 215; definition of, 12–13; transfer payments to, 6

Carter, James E., 1, 68, 108, 109
Chaining, 4, 6, 7, 155, 157–58, 159, 160, 166, 220; alternatives to, 164; on archaeological sites, 162, 167; costs and benefits of, 155, 162, 165; definition of, 156; justification for, 159; location of, 156; and wildlife, 163. *See also* Clear-cuts
Chee Dodge, 89
Civilian Conservation Corps, 89
Civil Service Reform Act, 18
Classification and Multiple Use Act, 124

Clear-cuts, 11, 158, 165, 186, 198.
    See also Chaining
Coal gasification, 4, 71, 72, 77, 78, 82
Coase theorem, 206
Cogeneration, 67
Collective action, 7, 8, 23, 46, 50, 83,
    84, 212
Collective control, 207, 211
Collective management, 5, 212, 213
Collective ownership, 33, 211, 212,
    214, 215, 220
Collective rights, 204
Collier, John, 89, 90, 104 n.8
Collusion, 26
Colorado Open Space Council, 187,
    194
Commission on Civil Rights, 102
Common pool resources, 3, 6, 13, 14,
    15, 38, 120, 143
Common property, 49, 51, 127
Common property resources, 47, 59,
    62 n.3, 74, 103, 175, 184 n.6
Commons, 13, 16, 149 n.19; demands
    on, 16; exploitation of, 14; logic
    of, 16; tragedy of, 4, 16, 127,
    149 n.19
Competition, 51, 62 n.3, 65, 211
Conservation, 47, 48, 49, 50, 51, 55,
    67, 68, 70, 71, 78, 109, 174, 212
Conservation Foundation, 1
Conservation movements, 29, 40, 42,
    125
Constitution, United States, 37, 38
Consumers, 16, 34, 50, 51, 53, 70, 77,
    78, 82, 209, 222
Continuous grazing, 131, 132. *See
    also* Grazing schemes
Contracts, 37, 38, 40, 49, 85 n.8
Corporations, 37, 40, 56
Cost-benefit analysis. *See* Benefit-cost
    analysis
Cost-effectiveness analysis, 166, 180,
    220
Cost efficiency, 16

Cost pricing, 78
Costs, 2, 4, 16, 57, 108, 114, 115, 116,
    119, 120, 122, 173, 197, 217,
    219, 221
Council on Environmental Quality
    (CEQ), 125, 147
Cross fencing, 101
Cross subsidy, 81, 83, 178, 184 n.8
Customary use areas, 88, 89, 90, 93,
    97, 99, 100

Decentralization, 60
Deferred rotation grazing, 130, 131,
    134. *See also* Grazing schemes
Definition and enforcement activity,
    22, 25, 28, 31, 33, 34, 35
Deforestation, 158
Department of Natural Resources,
    180
Deregulation, 69, 74, 78
Desert Land Act, 29
Due process, 37, 53
Duopoly, 10

Economic goods, 54
Economies of scale, 114, 115
Efficiency, 4, 22, 27, 28, 49, 50, 54,
    55, 61, 67, 72, 73, 78, 80, 84, 90,
    119, 170, 186, 187, 194, 200,
    201, 205, 213, 218; costs of, 191;
    economic, 6, 7, 144; gains in, 35;
    in government, 108; incentives
    for, 53
Efficiency tests, 192–93, 198, 199
Effluent, 52, 53, 54, 57, 58, 60, 61, 79
Effluent charges, 46, 48, 49, 57, 60,
    220
Elections, 9
Employment, 180, 221
Energy crisis, 77
Energy independence, 71
Energy policy, 66, 67, 68, 69, 218
Energy prices, 66, 72–73
Entitlement programs, 76 n.16, 84

Entrepreneur, 6, 20, 204
Environment: exploitation of, 51;
    impacts on, 58, 171, 179;
    policies concerning, 46, 47, 53,
    55, 57, 59, 219; quality of, 2, 46,
    50–51, 59, 77; regulation of, 56
Environmental crises, 22, 56
Environmental Defense Fund (EDF),
    68, 69, 78
Environmental degradation, 2, 4,
    6–7, 22, 49, 69, 76, 125, 179,
    186, 191, 199, 217, 220
Environmental hazards, 82, 83, 186,
    189
Environmental impact statements,
    124, 125, 137, 145, 146
Environmentalists, 1, 2, 6, 7, 48, 78,
    108, 121, 145, 163, 181, 187,
    191, 209, 219, 221
Environmental movement, 2, 156
Environmental programs, 109, 114,
    218, 221
Environmental Protection Agency
    (EPA), 51, 52, 55, 56
Equity, 26, 35, 49, 50, 90, 93, 100,
    102, 203, 204, 205, 206, 215,
    220, 223
Erosion, 87
Evaluation, 118, 119, 121, 122, 166,
    170, 171; of benefits and costs,
    119; of range land, 129; of
    water projects, 120
Even-flow constraint, 176, 178
Exclusion principle, 207
Exclusive rights, 25
Externalities, 3, 10, 41, 83, 84, 207

Federal Energy Regulatory Commis-
    sion, 79
Federal Land Policy Management
    Act, 124, 128, 133, 138, 143,
    147
Federal Power Commission, 4, 79
Federal Tax Code, 67

Fencing, 88, 99, 100, 101, 125, 130,
    136, 139, 194; agreements on,
    100–101, 102; incentive to, 100;
    on the range, 140–41
Field, Stephen J., 37
Fifth Amendment, 37
Flood control, 110, 112–13, 118
Flood Control Act, 113
Ford, Gerald R., 68
Forest: investments in, 176, 213;
    management of, 171, 175, 176;
    products from, 181; resources
    of, 171, 173, 174, 187, 193, 198
Foresters, 173, 174, 176, 178, 182,
    184 n.4, 184 n.8, 214
Forest Reserve Act, 126
Forest Service, 6, 40, 126, 130,
    149 n.19, 154, 155, 156, 159,
    161, 162, 163, 166, 167, 172,
    179, 187, 189, 190, 191, 195,
    198, 199, 201, 214, 219, 220;
    budget of, 178, 183, 191, 196;
    environmental impact state-
    ments of, 157, 159, 160, 162,
    163, 164, 165; justification for,
    188; management of, 187, 193;
    mandate of, 180; practices of,
    186, 197; responsibility of, 170
Fourteenth Amendment, 37
Freedom, 7, 36, 50, 84, 146
Free market economics, 1, 190, 193,
    199
Free rider, 3, 10
Frontier, 23, 24, 27, 28, 30, 36,
    38, 39
Frontiersmen, 33–34

Gadsden Purchase of 1853, 125
General Accounting Office (GAO),
    54
General Land Office, 126, 128. *See
    also* Bureau of Land Manage-
    ment
Government growth, 11–12

Government power, 10, 16, 27, 30, 38, 40, 53
Graduation Act, 34
Grazing, 40, 88, 89, 93, 97, 102, 124, 127, 128, 130, 133, 134, 138, 140, 143, 147, 155, 159, 223
Grazing committees, 97, 99
Grazing fees, 141, 144
Grazing management, 125, 134, 136, 144–45, 146, 219
Grazing permits, 93, 95, 97, 128, 129, 136, 143, 144
Grazing regulations, 95, 96, 97, 102
Grazing schemes, 124, 130, 133, 136, 137, 139, 143, 146; beneficiaries of, 141; costs and benefits of, 136, 141, 142, 146; effects of, 134, 137, 138; on public land, 134; purpose of, 146. *See also* Continuous grazing; Deferred rotation grazing; Rest rotation grazing
Grazing Service, 128, 149 n.19, 156

Hardin, Garrett, 4, 13, 14, 15, 17, 127
Homestead Acts, 29, 34, 35, 44 n.22, 126, 128
Homesteading, 33, 35, 43 n.9, 126
Hormay, August (Gus), 131, 132, 135

Ickes, Harold L., 89, 128
Incentives, 2, 3, 5, 6, 14, 18, 19, 24, 26–28, 31, 34, 43 n.9, 50, 53, 54, 55, 59, 60, 61, 64, 66, 73, 79, 81, 84, 99, 105, 127, 128, 147, 171, 174, 205, 207, 210, 212, 213, 215, 218, 219
Incentive structures, 13, 15, 16, 17, 19, 20, 53, 54, 58, 220, 221, 223
Indian Service, 87, 88, 90, 93. *See also* Bureau of Indian Affairs
Inefficiency, 22–23, 49, 57, 67, 83, 223
Inflation, 16, 18, 69, 71, 129, 219

Innovation, 36, 50, 55, 58, 59, 60, 65, 66, 214
Institutions, 1, 5, 30, 58; building of, 56; failure of, 121; reform of, 2, 183
Interstate commerce, 37, 66, 110, 209
Invisible hand, 10

Keynes, John Maynard, 204
Knutson-Vandenberg Act, 192, 193, 196
Kristol, Irving, 203–4

Labor, 23, 24, 28, 35, 200
Labor legislation, 38
Land: clubs, 30–31; disputes, 93, 95, 97; grants, 29, 126; laws, 29, 34; management units, 7, 101; policy, 34; rushes, 35
Land use planning, 113
Liquified natural gas (LNG), 68, 70, 71, 72
Livestock, 115, 125, 136, 137, 143, 164, 165, 166; and chaining, 160; response of, to grazing schemes, 133, 135, 136
Livestock industry, 128, 130, 132, 142, 143
Lobbyists, 178
Logging, 173, 188, 190, 195; aesthetic, 179; costs of, 179, 180; practices, 172, 191, 195
Louisiana Purchase of 1803, 125
Lurgi process, 78, 79, 82

Management: costs of, 189, 198; objectives of, 147; techniques of, 146
Market clearing price, 40
Market economy, 50, 173, 174
Market failure, 3, 7, 10, 11, 47, 188
Market intervention, 199
Market mechanisms, 206, 207

Marketplace, 171, 172, 174, 175
Market prices, 210, 217, 222
Markets, 38, 49, 50, 72, 119
Market system, 2, 9, 147, 183,
    184 n.6, 207, 213
Marshall, John, 37
Mining, 29, 31, 130, 158, 188, 199
Mississippi River Commission, 110
Missouri Valley Authority, 112
Monopoly, 10, 22, 37, 67, 82, 83, 84,
    211
Morgan, J. C., 89
Muller v. *Oregon* (1908), 38
Multiple use, 124, 128, 138, 140, 146,
    166; categories of, 134; defini-
    tion of, 133; objectives of, 140
Multiple Use-Sustained Yield Act,
    192
Munn v. *Illinois* (1877), 37

National Environmental Policy Act,
    157
National Forest Management Act,
    192
National forests, 11, 40, 126, 129,
    155, 159, 170, 172, 174, 178,
    183, 188
National Resource Defense Council
    (NRDC), 125, 135
National Wild and Scenic Rivers Act,
    109
Natural gas, 4, 64, 66, 67, 72, 74, 81,
    220, 223; allocation of, 71; al-
    ternatives to, 70; consumption
    of, 72, 73, 74, 78, 79; markets
    for, 72, 83; prices of, 70, 73, 78,
    80, 82, 83, 220; shortage of, 79,
    81, 83, 220; social costs of, 220
Natural resources, 2, 3, 109, 110,
    205, 211, 217
Negotiation, 88, 93, 100, 101, 206,
    219
Nixon, Richard M., 68
Nozick, Robert, 204

Office of Indian Affairs, 126
Office of Management and Budget
    (OMB), 114, 196
Oil imports, 68–69, 77
Opportunity cost, 6, 10, 11, 24, 68,
    71, 73, 84, 170
Ordinance of 1785, 29, 33
Organic Act. *See* Federal Land Policy
    Management Act
Overgrazing, 87, 96, 97, 99, 102, 127,
    128, 145, 157, 219; definition
    of, 87; explanation of, 102; in-
    creases in, 97, 99, 127; occur-
    rence of, 88

Pareto optimum, 206
Permits, 55, 90, 95, 100; national sys-
    tem of, 52; rationing of, 144;
    size of, 97; value of, 144
Pest epidemics, 163
Pick-Sloan Plan, 112, 116, 118, 119
Pinchot, Gifford, 126
Policy analysis, 204
Pollution, 46, 47, 56, 81, 179; abate-
    ment of, 48, 51, 52, 54, 57, 58;
    causes of, 50; costs of, 49, 50
Powell, John Wesley, 111
Predatory bureaucracy, 13, 17
Preemption Act, 29
Preservation, 208, 209, 210, 211
Price competition, 71
Price controls, 82, 220
Prices, 40, 147, 215, 221
Pricing mechanism, 215, 220
Pricing policy, 2, 66, 77, 221
Prior appropriations doctrine, 33
Prisoner dilemma, 26, 43 n.11
Private benefits, 10, 23
Private costs, 10, 23, 49
Private enterprise, 29, 35, 37
Private property, 3, 38, 84, 126, 129,
    174, 212, 217, 219
Private rights, 26, 27, 33, 36, 38, 204
Production process, 47

Productive activity, 11–12, 24, 28, 43 n.15, 128, 205
Profit motive, 2, 5, 217
Profits, 3, 15, 37, 50, 170, 173, 192
Progressive movement, 40–42
Project development, 121
Property rights, 2, 3, 22, 26, 34, 49, 51, 87, 99, 127, 144, 204, 206, 210, 211, 213, 215; definition and enforcement of, 23, 27, 34, 46; economic theory of, 23, 88; establishment of, 28, 38; evolution of, 23; in natural resources, 3, 206
Property rights paradigm, 5, 7
Public choice, 5, 49
Public domain, 22, 29, 33, 34–35, 36, 40, 46, 129
Public goods, 3, 10, 21 n.2, 84, 116, 172
Public interest, 13, 19, 20, 37, 171, 180, 213, 217
Public land, 29, 125, 126, 133, 146
Public Land Law Review Commission (PLLRC), 129, 130, 143
Public management, 49, 213, 218
Public range, 146
Pure Food and Drug Act, 172

Quine, W. V., 204

Railroads, 35, 188, 199
Range demonstration area, 95
Range land, 32, 88, 125, 127, 130, 147–48, 159, 220; administration of, 128, 160; benefits from, 148; carrying capacity of, 87, 89, 90, 96, 97, 102, 130; condition of, 89, 127, 129, 130, 134, 135, 143; rights to, 32, 33
Range management, 90, 101, 129, 141, 145
Rational ignorance, 6
Rawls, John, 204

Reclamation, 54, 119
Reclamation Act, 110
Recreation, 11, 110, 114, 115, 125, 130, 133, 139–40, 161, 166, 172, 181, 186, 187, 188, 190, 199, 200, 201
Reforestation, 173, 187
Regeneration, 188, 193, 194, 196, 198
Regulation, 16, 53, 56, 65, 172
Rent dissipation, 24, 25, 26–27, 28, 30, 31, 32, 33, 34, 35, 38, 40
Rents, 24–27, 43
Rent-seeking, 22, 23, 36
Residual claimants, 27, 33, 217
Residuals, 26, 28, 47, 48–49, 188, 189
Resources, 24, 65, 87, 113, 205, 209, 212, 215; allocation of, 14, 22, 35, 36, 40, 46, 51, 53; exploitation of, 24, 207; prices of, 7, 10, 23; rights to, 27, 30, 38; scarcity of, 58, 59, 205; transfer of, 207, 214; waste of, 54
Rest rotation grazing, 124, 130, 131, 132, 134, 135, 136, 139, 140. *See also* Grazing schemes
Risk aversion, 15, 100
Road building, 178, 179, 191, 192, 193, 194, 196
Rolled-in pricing, 69, 73, 74, 80, 81, 83, 84
Roosevelt, Franklin D., 111, 112
Roosevelt, Theodore, 126
Rule of reasonableness, 37–38
Rule of willing consent, 3, 7, 17

Scarcities, 14, 58, 59
Self-interest, 2, 5, 15, 17, 19, 20, 121, 204, 205, 208, 209, 210–11, 212, 213, 217
Slaughterhouse Cases (1973), 37
Smith, Adam, 9
Smith, Jared, 131

Social benefits, 3, 10, 19, 20, 23, 81, 121

Social costs, 1, 10, 11, 13, 20, 23, 49, 108, 121, 171–72, 200

Social waste, 27, 42, 54

Society for Range Management, 129

Soil Conservation Service (SCS), 109, 110, 111

Solar energy, 67, 73

Special interest groups, 4, 6, 11, 17, 41, 68, 119, 171, 222

Speculators, 31, 209, 210, 211–12

Spillover benefits, 16

Squatters, 29, 31, 32, 33

Stock reduction program, 87, 89, 90, 95

Stumpage prices, 188, 194, 195, 196, 197, 198, 199, 200

Subsidies, 2, 6, 7, 52, 54, 65, 66, 67, 69, 75 n.8, 109, 117, 143, 147, 189

Supreme Court, United States, 36–37, 40, 82, 110

Sustained yield, 124, 128, 133, 159–60, 173, 188, 195

Sustained yield-even flow, 171, 172, 173, 174–75, 176, 178, 179, 181, 182, 183, 220

Swayne, Noah H., 37

Synthane, 82

Synthetic natural gas (SNG), 67, 68, 69, 72, 77, 79, 80, 81, 220; commercialization of, 83; costs of, 69, 70, 71, 78, 82; dangers of, 79; demand for, 83, efficiency of, 81; production of, 79, 84; subsidies to, 80, 81, 83; success of, 80

Tax Reform Act, 52

Taxes, 5, 6, 54, 61, 64, 65, 66, 67, 68, 72, 73, 74, 116, 126, 220, 221

Taxpayers, 6, 12, 16, 51, 117, 171, 180, 218, 222

Taylor Grazing Act, 89, 128, 144

Technical assistance, 109, 110

Technological adaptation, 56, 58, 59, 60

Technology, 47, 50, 55, 56, 61, 71, 74, 81, 207

Tennessee Valley Authority, 108, 111, 112. *See also* Water development agencies

Third party effects, 49, 50

Timber, 6, 11, 29, 35, 40, 174, 180, 182, 183, 187, 191, 192, 193, 194, 195, 197, 212, 214, 221; demand for, 181; exploitation of, 126; harvesting of, 171, 172, 175, 179, 180, 189, 190, 191, 192, 194, 219; management of, 180, 186, 190, 191, 192, 194, 195, 196, 197, 198, 199, 200; market for, 181, 186, 188, 189, 191, 200; private ownership of, 171; sale of, 170, 178, 182; scarcity of, 173, 174, 175, 188; supplies of, 173, 174, 181

Timber Culture Act, 29

Timber industry lobbyists, 178

Timber mining, 192, 202 n.9

Timber Stone Act, 29

Transaction costs, 3, 95, 100, 102

Transfer activity, 11–12, 24, 27, 28, 30, 39, 40, 42, 93, 144; magnitude of, 40–41; in nineteenth century, 36; prevention of, 41; resources used in, 38

Transfer society, 9, 11, 12, 41

Transformation process, 47, 48

Transgenerational equity, 204, 205, 206, 215

Transitions, 51, 55, 60, 61

Treasury, 12, 13, 14, 15, 16, 84, 125, 147

Tribal Council (Navajo), 88, 89, 90, 95, 99, 102, 219

Tribal Resources Committee, 100

Turner, Frederick Jackson, 23, 27, 30

Unemployment, 18, 69, 182, 200, 201, 218
Union of Radical Political Economists, 204
Urbanization, 115, 119
Usufruct rights, 88, 102
Utilities, public, 67, 68, 69, 71, 74, 80–81

Waite, Morrison R., 37
Waste, 17, 23, 24, 27, 53, 54, 67, 122, 178, 214
Water development agencies, 108, 114, 115, 119, 121. *See also* Army Corps of Engineers; Bureau of Reclamation; Tennessee Valley Authority
Water developments, 4, 108, 109, 110, 111, 114, 115, 117, 121, 136, 219, 223; construction costs of, 117; cost-benefit analysis of, 117; environmental problems of, 108, 118, 120; evaluation of, 116; justification of, 113, 114; social costs of, 116, 118, 120

Water law, 33
Water policy, 109
Water pollution, 52, 53, 55
Water Pollution Control Act, 51; amendments to, 51–52
Water quality, 52, 53, 54
Water resources, 115, 119, 192
Water Resources Council, 118, 196
Water rights, 32, 33
Wealth distribution, 26
Wealth maximization, 102
Welfare, 57, 93, 102, 116, 206, 211
Welfare triangle, 10, 22
Wellhead price regulation, 4, 68, 71, 79, 80, 81, 83–84, 85 n.8
Wilderness areas, 172, 186, 188, 189, 191, 197, 199, 200, 223
Wilderness Society, 187, 194
Wildlife, 114, 115, 116, 120, 125, 130, 137, 139, 142, 147, 164, 166, 174, 184 n.6, 192; and chaining, 161–62; exploitation of, 174; and fencing, 137, 139; and grazing schemes, 133, 138, 139
Windfall profits, 78

DATE DUE